Managing Health and Safety
in Bui

Managing Health and Safety in Building and Construction

Tony Clarke, FCIOB, MCIPS, FFB, registered adjudicator

OXFORD   AUCKLAND   BOSTON   JOHANNESBURG   MELBOURNE   NEW DELHI

Butterworth-Heinemann
Linacre House, Jordan Hill, Oxford OX2 8DP
225 Wildwood Avenue, Woburn, MA 01801-2041
A division of Reed Educational and Professional Publishing Ltd

 A member of the Reed Elsevier plc group

First published 1999

British Library Cataloguing in Publication Data
Clarke, Tony
  Managing health and safety in building and construction
  1. Construction industry – Employees – Health and hygiene
  2. Construction industry – Safety regulations – Great Britain
  I. Title
  363.1'19624'0941

ISBN 0 7506 4015 4

Library of Congress Cataloguing in Publication Data
Clarke, Tony.
  Managing health and safety in building and construction/Tony Clarke.
  p.   cm.
  Includes index.
  ISBN 0 7506 4015 4
  1. Construction industry – Safety regulations – Great Britain.
  I.   Title.
  KD3172.C6C58.   99–36658
  343.41'07869–dc21   CIP

Composition by Genesis Typesetting, Rochester, Kent
Printed and bound in Great Britain by Biddles Ltd, Guildford and King's Lynn

Contents

Preface

My career in construction began with a sort of false dawn in 1966 when I started work as a labourer on a housing site in my school holidays. My only attributes were that I was lively, fit and big for my age, and willing to have a go at most things. My willingness was soon put to the test and I quickly found myself knocking up mortar and hod-carrying for the bricklayers, loading up plasterboard, two sheets at a time, for the plasterers and doing a variety of tasks including jack hammering out other people's mistakes, driving dumpers with no synchromesh – I had not yet passed my driving test – and acting as chain man to the site engineer. By the end of my holiday I was using his quick-set level and theodolite while he went down to the bookies. I loved it. The money was good, about £15 a week as I recall, I got a splendid suntan from the waist up and the work kept me fit for my real passion in life which was rugby.

Further sixth form and college summers came and went in the same way. It was easy to get work and in truth, you could do as much or as little as you pleased. It is in my nature to get 'stuck in', so I found myself one summer in a heavy gang working 12 hour shifts pouring concrete non-stop in a large cut and cover tunnel. Because I was a student, the hours I worked were not set against the bonus target so in a six-man gang we were nearly 20 per cent ahead of target before we had started. It was a matter of pride to me to pull my weight in the gang so I would toil away with my shovel, knee deep in concrete, balancing on steel reinforcement bars that I couldn't see, curved on plan and elevation, until I realized one day that the rest of the gang did it all with vibrating pokers and never set foot in the sticky stuff. That is why I got dermatitis on my legs that year, and they did not.

In July 1970, I started work as a management trainee with a major national contractor in the north west of England. Over the next six or seven years I worked on a range of building and civil engineering projects including an oil refinery, a power station, a large office block, more housing and a couple of motorways. Life was never dull and the cast of characters met, friendships made and experience gained could provide enough material for a TV soap opera. Time was always at a premium, and there always seemed to be some desperate hurry to start a new job or finish an existing one.

Every now and then, the company safety officer would visit, or there would be a safety roadshow, but to my shame, these always seemed to be impositions and intrusions into other activities that had higher priority in my mind.

In 1970 I joined a new company with a new job on a new motorway in a new town in East Anglia with a newish family with a new salary and a title that included the word 'manager' in it somewhere. It was sobering to realize that having made it half way up the greasy pole, and taking into account the company car, I was earning about half the wages I had made in my college summers.

After the longest hottest summer on record, we had the longest wettest winter and before the project was six months old, we were twelve months behind programme. People outside the construction industry are always mystified by this phenomenon. Construction professionals know it all too well. In the summer of 1977 the miserable weather broke at last and we were making some in-roads into the delays. We had also had some training on the relatively new Health and Safety at Work etc. Act, which had finally penetrated into my consciousness, partly because of the criminal penalties attaching to the law. Probably as a result of the increased awareness that extra responsibility brings, I was beginning to ask some rather awkward questions about the way things were done on the project, or at least, the half of it for which I was accountable.

Driving down the temporary haul road one bright afternoon in 1977, my radio crackled into life. It told me that there had been a trench collapse on an interchange at the eastern extremity of the site. Three men had been in the trench at the time, two had got out and the third was trapped up to his armpits. However he was conscious and talking and the section foreman was confident he would be released in a few minutes. I checked that the message had been received at base station and that accident and emergency procedures had been activated, and I drove straight to the accident site at the interchange. It took me about ten minutes. When I got there, the trapped operative was dead, killed by crushing injuries to his chest and abdomen even though he was only trapped up to his armpits.

Along with many of my colleagues I was profoundly affected by this incident for a long time afterwards. Nobody had done anything wrong or stupid; the accident happened as some trench supports were being installed, and that was the point as far as I was concerned, because it brought home to me the foolhardy risks that were taken on construction sites where far less safe work methods were employed as the norm.

From that time on, I took a far greater interest in health and safety but within the overall context of safe, efficient work systems. From 1980 to 1994 I owned and operated my own building and project management company working within a 50 mile radius of Cambridge where I live with my family.

I came to the view that health and safety is not, and never can be, some separate issue. It must take place within the process of construction, and for that reason I came to see the issue of the Management Regulations 1992 not as an imposition but an opportunity. The Approved Code of Practice accompanying the Management Regulations is in fact excellent business advice which, if implemented fully in any business, would result in a total quality management system all for the purchase price of £7.95. The Construction Industry Training Board (CITB) offer excellent training in the application of all relevant regulations for what amounts to bargain basement prices.

Another interesting thing about training was that motivation and productivity levels always rose significantly immediately after the training session or course.

Many of my local competitors, some of whom were and are also friends, with whom I was in regular contact, did not share my views. Most saw health and safety as an imposition. I was occasionally the butt of their humour for having more than a passing interest in and knowledge of the Construction Regulations, and was looked upon as an 'anorak' when I enrolled on several courses dealing with the 'six pack' in 1992.

I never minded this, because I believed it gave me a significant business advantage over them. They paid consultants considerable sums of money to give them both business advice and health and safety advice, that I felt I was obtaining free of charge, courtesy of the Health and Safety Executive in the form of Approved Codes of Practice and Guidance Notes.

The other thing I came to realize was that my competitors did not understand health and safety law. They felt that it was too complex, costly, time consuming and difficult to understand. The prudent ones delegated, and the imprudent ones ignored their obligations as employers. All of them had their in-house subcontracts and supplier agreements altered by lawyers, usually at great expense, every time some new law was made. They were paying in-house safety officers and some of them also paid for external advice. It cost my competitors money they didn't need to spend to comply with the law, whilst I made money out of complying with it. Even so, they were still able to quote cheaper prices than me in competitive tender. I couldn't understand this, and I promised myself that one day I would write a book about it, if ever I got the time.

Along with everybody else in construction, 1993 was a bad year for me and my company. Clients and their lawyers discovered 'design and build', and were imposing onerous design and build conditions and suicidally low prices on contractors. Always a bit of a loner, I stayed out of this to a large extent, but it meant that my work load fell substantially. This gave me time to think and to reflect, and after discussions with my family I realized I did not want to see out the final third of my career in the adversarial battlefield that the construction industry had become.

We wound up the business just as the CDM Regulations came into force in 1994, and I joined James R Knowles Ltd, ranked fiftieth in the top 200 UK construction consultancy businesses. My specific task was to establish and manage a new division of the company offering a range of consultancy advice and CDM-related services including planning supervision. I was able to grow the business quickly, but was obviously limited by the dearth of competent planning supervisors. Also my old conviction that there was no such thing as health and safety was not a good marketing mantra to those who wanted to believe the opposite.

The interesting thing about CDM is that in order to do it properly, it makes duty holders, clients, designers, principal contractors and subcontractors take a long hard look at their own business and for the most part it forces people to invest in training, training, training.

In 1994, Sir Michael Latham published 'Constructing the Team', and I started a MSc in Construction procurement at the Nottingham Trent University.

Somehow, I have no idea how, all of these strands came together in 1996/7 when I expanded the Health and Safety Management Division of the Knowles Group into a new business embracing construction procurement and health and safety management strategies. I was also astonished and delighted to find myself working closely with Sir Michael, who had joined the Knowles Group in a non-executive capacity as part of the Group's flotation plans for 1998.

In addition to my full time business activities, I also began lecturing and training to people and organizations in issues arising from CDM, but found that most of them knew nothing of the HASWA/MHASWA requirements, philosophy and approach. I found I was able to call upon a rich vein of practical experience, which was shared with many of the people attending my lectures and training courses, but I was also able to relate this pool of business and technical experience to detailed knowledge of the main provisions of the more important sets of regulations applying to construction.

I became more and more convinced of the real need for a book such as this, and having completed my Masters I decided to chance my arm. I only hope that what seemed, and still seems to me, to be the genuine commitment of an experienced construction professional, does not instead appear to be the incoherent ramblings of a vainglorious scribbler. In other words, I hope you like the book.

Tony Clarke

1999

Acknowledgements

I would like to extend grateful thanks to Paula Sayer who did the substantive preparation of this manuscript and to Joan Orr for stepping into the breach to finish it off when Paula moved on to pastures new.

I would also like to thank Rebecca Hamersley, and everybody else at Butterworth-Heinemann for their endless patience whilst waiting for me to deliver the final draft of this manuscript.

I also owe Roger Knowles a huge debt of thanks in providing me with the opportunity and the platform, through the James R. Knowles Seminars and Training Division, to present my particular ideas on health and safety management in construction to a very wide range of organizations and individuals. I would like to thank Sir Michael Latham for, first, encouraging me to start the book, and second, encouraging me to keep going to finish it.

Finally I would like to thank my wife, Chris, and my children Adam, Laura and Abigail for tolerating my uncommunicative presence in the study over so many nights and weekends.

1

Brief history of health and safety law in the UK

1.1 Introduction

The first recorded health and safety law is thought to have been laid down by King Hammurabi of Babylon in 2200BC when it was said that,

> If a house shall fall down and kill the owner, the builder of that house shall be put to death. If a house shall fall down and kill the son of the owner, the son of the builder shall be put to death.

Some elementary building regulations are recorded in the *Bible*:

> Build ye thine houses with parapets so that blood shall not stain the stones of thine house if a man fall from the roof. (Deuteronomy).

In the UK the first recorded law was passed in 1802 and dealt with the health and morals of apprentices in the cotton industry. From that time, development of the law can be seen in two distinct and separate phases, between 1802 and 1974, and 1974 to the present.

The first period, 1802 to 1974, covers the development of the modern United Kingdom from a mediaeval, largely agricultural society and economy to what it was in 1974. For most of that time, and for the majority of the population, the UK was probably not a very pleasant place in which to live and work. Charles Dickens wrote some compelling stories chronicling the times. One of them, *Hard Times*, published in 1854, has been described by critics as 'an industrial novel, relating to northern industrial England in the mid 19th century, and the dreary, oppressed conditions of the workers. It also investigates the mindset of those who persist in seeing these workers as mere useful tools, as "hands . . ." '

A newspaper of the time carried the following article describing what was probably a typical industrial accident:

> There are many ways of dying. Perhaps it is not good when a factory girl, who has not had the whole spirit of play spun out of her for want of meadows, gambols upon balls of wool, a little too near the exposed machinery that is to work it up, and is immediately seized, and punished by the merciless machine that digs its shaft into her pinafore and hoists her up, tears out her left arm at the shoulder joint, breaks her right arm, and beats her on the head.

The article subsequently points out that the girl survived the accident.

In the proof of *Hard Times* Dickens includes a description of a similar accident which happened to one of his characters. The punchline was that the accident happened 'despite government directives because manufacturers regarded the boxing off of dangerous machinery as: "Unreasonable! Inconvenient! Troublesome!".' We can only guess at the reasons why Dickens did not subsequently include this passage in the print of the novel.

As industry developed, so too did health and safety law. In 1833 an Act was passed dealing with 'Factories' and in 1866 further legislation was made for 'Shops'.

The contents of these long repealed laws are unimportant now, but what is instructive is the approach they took. They were prescriptive and narrow, in the sense they told employers what to do in those particular workplaces. Thus on the one hand affected employers had the burden of complying with the law, but on the other, at least they knew the limits and extent of what they had to do. Similarly, their employees, if they were literate, had a clear idea of their rights, such as they were.

However, this approach meant that many industries, workplaces, employers and employees, were outside the remit of any sort of health and safety law. Factory owners were expected to achieve one set of standards, shop owners another, mill owners another and so on. This lead to the development of a 'patchwork quilt' of laws containing many inequalities, which was barely understood by the enforcers, let alone those whose duty it was to comply.

A point to remember is that illiteracy was the norm, at least until the early part of this century, and must have had a significant impact on both the application of and compliance with the law.

The main industries covered by the legal framework were factories, railways and shops, mining and steelworks. The effectiveness of both compliance and enforcement was, to say the least, questionable.

My father, who worked on the docks and in construction on Merseyside between the First and Second World Wars, told me chilling tales of the conditions of the time. On the docks, men queued in pens every morning from 06:00 onwards and if selected, worked that day and were paid subject to a levy made by the 'walking pelter' or gangmaster who selected them. They could then be put to work for anywhere between 4 and 16 hours at a time in the hulls of ships unloading coal or grain or asbestos or iron ore or some other cargo, by hand, shirtless, in unlit, unventilated spaces on a gang price per ton. Consequently, nobody stopped to eat or drink, to wash or to go to the lavatory, even if such facilities were available.

In construction, things were no better. 'Queensway', the first Mersey tunnel from Birkenhead to Liverpool, was opened in 1922. Over 70 men were killed and several hundred injured working on it over the five years it took to build. These figures would provoke a national outcry now, and more prosaically, would attract HSE enforcement action.

Against this background, in the early 1970s the government of the day invited Lord Robens to undertake a complete review of the health and safety system in the UK, in much the same way as Sir Michael Latham reviewed the UK construction industry more recently in 1994.

Lord Robens had previously been chairman of the National Coal Board and had successfully implemented change throughout the coalmining industry through an approach that put efficiency and safety at the heart of his reforms. Thus, the Robens Commission lead to the Health and Safety at Work etc. Act 1974 (HASWA) and the radically new approach envisaged by Robens. Basically, the new Act introduced the risk assessment/goal setting philosophy that underlies all modern health and safety law. It laid down the requirement that, essentially, all workplaces, whatever sector they are in, must achieve similar safe work conditions, that all employers must achieve the same standards of workplace and occupational safety and must treat employees with equality. Employees were given certain duties too.

In effect, this marked a 180° turn away from the old prescriptive 'tell them what to do' approach, to the new goal setting methods where employers and employees are told what they must achieve, given a methodology – risk assessment – for achieving it, and are then expected to achieve compliance through competence.

The Act was a watershed in UK health and safety laws in more ways than one. Apart from introducing the new approach, the Act was also intended to provide a framework for the repeal of old outdated laws and their replacement by newer modern laws written to match, as far as possible, the new risk assessment approach.

Over time, the intention was to repeal all pre-1974 laws and to simply have the Health and Safety at Work etc. Act 1974 on the statute books, accompanied by appropriate regulations, approved codes of practice and guidance, and since 1974, that is the direction that various governments have taken. At that time, the UK was not a member of the European Community but has since joined, and this has had a significant effect on health and safety law in this country. Whilst the overall intention to repeal all pre-1974 laws remains unchanged, successive governments have faced far more detail in the form of proposals for new laws flowing out of the EU for implementation in member states.

It is probably true to say that this has significantly slowed down the process of domestic reform for two reasons. First, governments have waited to see what Europe wanted before reforming domestic legislation. Second, Europe has asked for laws which the UK might not have introduced of its own accord. Regrettably, in the UK this has lead to a situation where, since 1974, many old prescriptive laws have sat alongside the new risk assessment regime and there is clear evidence that business and industry find this a confusing situation.

So the present position is that the HASWA philosophy prevails and risk assessment is the main engine of health and safety management in the UK. However, employers and employees must be mindful that many old style prescriptive laws are still in place either fully, or worse, partly. Events and developments in the EU also impact directly on the UK legal framework, and are seen to influence many new regulations issued here.

1.2 The cost of accidents

No one knows the true cost of accidents.

There is a plethora of statistics about accidents and costs but they are fraught with difficulty. Any researcher soon learns that the basic problem arises from:

- lack of research into accidents
- the way in which accidents are reported
- the unreliability of data
- lack of agreement about what is meant by an accident

Taking all of these matters into account it appears that accidents cost Great Britain between 1.75 and 2.75 per cent of its annual turnover.

In the ten year period 1986/7 to 1996/7, fatal injury rates to workers at work were consistent at around 1 to 1.5 per 100 000 workers, equating to between 300 and 450 deaths per year. In addition, around 158 000 non-fatal injuries are reported each year, and an estimated 2.2 million suffer ill health because of work. By way of contrast, around 4500 people are killed each year on our roads and around 4200 are killed in domestic accidents.

Agriculture and construction both have traditionally poor safety records in comparison with other industries. Agriculture consistently records fatality rates of between 8 and 11 workers per 100 000 and construction between 5 and 6 workers per 100 000.

One of the main reasons put forward by experienced commentators for these figures is the extensive and increasing use of self-employment in these sectors in recent times, particularly in construction where labour-only subcontracting has been the norm since the mid-1980s. It is axiomatic that self employed persons are often used at the workface and may not have as much training, experience and managerial support as they should have.

An illustration of the dilemma in calculating the cost of accidents was given by the Government Safety Minister, Angela Eagle, in 1997. In response to a parliamentary question on the true cost of accidents, she said that in 1990 – the latest year that full figures were available at that time – the total cost to the British economy was between £6 and £12 billion per annum. When pressed, she added that if human pain, grief and suffering were taking into account, the figures rose to £11 to £16 billion.

The uncertainty arises from the way in which costs are calculated. Usually this is done under two main headings:

■ insured, and
■ direct

But this takes no account of two further headings under which costs can be incurred.

■ uninsured, and
■ indirect

In the *Piper Alpha* explosion, 167 lives were lost and £746 m was paid out by insurers, but estimates put the total loss, including business interruption, investigation costs, hiring and training replacement personnel and the like, at over £2 billion.

5

Insured and indirect costs of the explosion include:

- Employer's liability
- public/third party liability
- all or combined risks
- property and material damage
- business interruption
- motor vehicles
- corporate liability.

Uninsured and indirect costs of the disaster include:

- product and material damage
- plant and buildings damage
- tool and equipment damage
- legal costs
- emergency purchases
- cleaning and clearance costs
- overtime working
- temporary labour
- investigation time
- management diversion time
- loss of expertise/experience
- loss of business, image and goodwill
- criminal penalties.

Accidents are generally classified as follows:

- over-three-day injury accidents
- minor injuries
- non-injury accidents.

It is this combination of insured/uninsured/direct/indirect/over-three-day/ minor injury/non-injury criteria that results in unreliable data, for there is no central databank into which all of this information is placed, collated, analysed and interpreted.

There is a statutory requirement to report over-three-day accidents, but not the cost flowing from them. There are statutory requirements and contractual obligations arising out of insurance arrangements but these can sometimes take years to resolve, and this in large part explains why there is such a time lag in obtaining cost data. For example, an insurance claim for compensation for injury arising from an accident might typically take 2–3 years to resolve.

A similar claim for occupational health or disease-related compensation may take decades to come to light, and, if disputed, 5–7 years to resolve.

These issues are reflected in the apparent vagueness and broad generality of the statistics. There is a vast difference between the low figure of £6 billion and the high figure of £16 billion quoted by Ms Eagle in her reply.

A more realistic, if not more accurate approach to assessing the cost of accidents was taken by the Accident Prevention Unit of the Health and Safety Executive in 1989. They ran five case studies which produced results which could be compared within operational parameters, and which are broadly summarized below.

Accidents cost:

- a construction company 8.5 per cent of its tender price
- a dairy product manufacturer 1.4 per cent of operating costs
- a transport company 37 per cent of annualized profits
- a North Sea oil producer the annualized equivalent of £3.76 m
- an NHS hospital 5 per cent of annual running costs

For every pound of insured costs recovered, the case study participants lost between £8 and £36 in uninsured, unrecovered costs.

In each case, an appropriate definition of accident was made. For example, on the construction site an accident was defined as: 'any unplanned event resulting in material loss or damage, or injury, and which costs more than £5'.

Clearly, this extended to such incidents as a pallet of bricks being pushed over and made useless, and in total 3626 such incidents were recorded in the 18-week period of the case study; 56 of these involved minor injuries, 3570 were property damage – in other words, waste caused by careless work habits and practices. There were no over-three-day or serious injury accidents recorded.

Closer scrutiny of the case studies as shown in the comparative figures in Table 1.1 is instructive.

Table 1.1 Construction accidents compared

	Construction	Creamery	Hospital	Oil	All
Injury:non-injury	1:64	1:24	1:18	1:25	1:37
Accidents/year/employee	87	11	7	5	17
Insured/uninsured	1:11	1:36	n/a	1:11	n/a

In broad terms, the implications are striking.

- Most accidents do not involve injury or ill health.
- Most accidents are preventable.
- Most accidents are a complete waste of money.
- Construction seems to have about three times as many preventable accidents as other industries.

Awareness of these matters has been growing rapidly across all sectors of business and industry, and was given particular impetus by the introduction of the 'six pack' in 1993, and in particular the Management of Health and Safety at Work Regulations 1992 and Approved Code of Practice.

Since the mid-1970s, some industrialists and health and safety professionals have advocated an approach described as 'total loss control', arguing that health and safety is indistinguishable from good business practice, and health and safety management arrangements are both ethical and of assistance in reducing organizations' costs.

Most line managers know that accidents cost money and if they are prevented, profitability can improve. But it costs time, trouble and money to prevent them, initially at least, in the form of improved education, training, control procedures, management commitment and the employment of better calibre people.

Because the exact extent of preventable loss is never easy to define, it is difficult to make out a business case or an actuarial case to substantiate spending on loss control. Those opposed to taking specific measures point out that it can cost more money to implement loss control than is saved by it.

1.3 Criminal and civil law explained

It is important to differentiate civil and criminal law because all health and safety law is made under the criminal code and carries with it criminal penalties for breach. There are three branches of UK law:

- common law
- civil code
- criminal code

The common law is judge-made law which comes down to us in the form of case law and precedent set by the judgments of the courts. It plays a diminishing role in health and safety due to the increasing volume of written laws made in modern times.

Civil and criminal laws are both made by parliament and are the law of the land. Where criminal law is broken a crime is committed. The state enforces the criminal code and punishes those who break it by means of fines and imprisonment.

Where civil law is broken, it is generally by way of breach of a contractual obligation, and whilst one or more of the parties to the contract may have suffered loss and expense arising from the breach, there are unlikely to be any criminal consequences or physical injuries arising, and the parties must enforce the provisions of their agreement themselves.

A contract written in the civil code cannot be written in such a manner as to oppose any provisions of the criminal code.

A successful action under one code can often lead to further action under the other. If an accident happens at a workplace, it is not uncommon for there to be a prosecution under the criminal code for failure to comply with statutory law followed by a separate action for negligence or breach of contract under civil law.

The burden of proof required for a successful action under criminal law is usually much greater than under civil law. If the first action succeeds, then the second one is almost certain to be successful.

If an accident happens then the person responsible can be prosecuted and if found guilty, fined and jailed. The person can then be open to an action for damages by the injured party or relatives of the party and may be liable for substantial damages which would have to be met out of his or her own pocket.

An illustration of the point is given by an accident that occurred in Derby in 1989. A passenger on a bus was killed when an excavator slewed and hit the top deck of the bus as it was passing through the site of a road resurfacing contract. The client, Derby City Council, together with the main contractor and one of its subcontractors were all charged with offences against the Health and Safety at Work Act. The road resurfacing works were let under a form of contract known as the ICE 5th edition.

Within the matrix of the contract, Derby City Council was the employer and also used its own staff to undertake the professional duties of the engineer and the resident engineer. The main contractor was employed to carry out all of the road resurfacing works which included temporary works in the form of various diversions and a contra flow. The subcontractor was employed by the main contractor to provide labour and plant to assist in the ripping up and disposal of the old carriageway surfacing.

The main contractor and subcontractor had devised the work method in the contra flow where the digger, which was a tracked backacter, ripped up the carriageway, scooped up the arisings in its bucket and then slewed through a 180° arc across the closed-off section of carriageway to load the disposal wagons which were backed up behind it. For part of the contra flow, a line of trees growing in the pavement alongside the works prevented the digger from slewing in the in-board direction across the closed-off carriageway. So the digger driver slewed in the opposite direction across the live carriageway and the digger bucket hit the top deck of the bus as a result of this work method.

At first, all three parties defended their actions. Derby City Council pointed out that it had relied upon the terms of the ICE conditions of contract which places safety responsibility with the main contractor.

A Health and Safety Executive spokesman said:

> We don't enforce ICE Contract Rules, we enforce the law of the land. If we think that an accident of this type has been brought about by more than one party, then we will continue to take action against more than one party. There is nothing in the Health & Safety at Work Act which says that health and safety is the responsibility of only one party.

Subsequently all three parties pleaded guilty to the offences with which they were charged and all were convicted and fined. The council and the subcontractor were each fined £15 000 and the main contractor was fined £5000.

At Derby Crown Court, Judge Brian Woods said:

> digging up roads has become almost a national pastime and it is the responsibility of the diggers and their employers to ensure that work is done without risk to the public.

The lessons to be learned from the judgment are:

- Safety can be the responsibility of more than one party.
- The council was found liable because its engineers were either resident or had visited the site on a number of occasions.
- They had not noticed or acted upon unsafe working practices.
- The main contractor was probably not directly responsible for the accident but he too could have taken action to prevent it.
- The subcontractor was directly responsible for the accident, and should have prevented it.
- Terms written into a contract provided no defence to the person who wrote the contract.

Prior to this case, the general view throughout business and industry was that because the head contractor in any contractual arrangement has full civil contractual liability for health and safety matters within the contractual matrix, then he also attracts full criminal liability, and that professionals need not concern themselves with health, and in particular safety issues, other than through writing lengthy, verbose and often irrelevant insertions into the head contract terms.

The case clearly demonstrated that this notion is quite wrong. However, experience shows that the lesson has not been widely learned, even though this test case occurred in 1988 and there have been many similar subsequent cases reinforcing the message. Many contractual arrangements still deal with health and safety in a similar manner.

Approaches to dealing with health and safety in contract are discussed in greater detail in Chapter 9.

1.4 Acts, regulations and ACoPs explained

The hierarchy of health and safety law is as follows:

- Acts of Parliament
- Regulations
- Approved Codes of Practice
- Guidance

Acts of Parliament

Acts are made by Parliament and they are the law of the land.

They place general duties on employers and employees and usually apply to all work places and they amount to a statutory enactment of common law 'duties of care'. They generally place three levels of duties on dutyholders, but in general dutyholders are required to do only what is 'reasonably practicable' to control a risk and this envisages that in assessing control measures some kind of cost benefit analysis will be carried out.

Regulations

Usually, Acts simply set out the requirements of the law and cannot be implemented unless accompanied by a set of Regulations. These do not normally add to the scope of the general duties imposed by the Act. Instead, they set out the minimum standards which must be achieved. Sometimes they may impose a higher standard of duty than that set by the Act, for example, by imposing an absolute duty in place of a reasonably practicable one. It is usual for Acts to give ministers the power to issue Regulations needed to implement the law.

Approved codes of practice

Approved Codes of Practice (ACoPs) can sometimes accompany Regulations and the intent is to clarify particular aspects of the general duties. If Regulations set out the minimum requirements of the law, then ACoPs give guidance on what must be done to comply with the law.

An ACoP is not law but it has the force of law. Therefore, failure to comply with any provision of an ACoP is not in itself an offence but that failure may be taken by a court in criminal proceedings as proof that a person has contravened the Regulations to which the provision relates. In such a case, however, it would be open to that person to satisfy a court that he or she has complied with the Regulations in some other way. In practice, it is usually easier to simply comply with the Approved Code of Practice.

Guidance

Guidance has no legal status. It is issued by HSE to assist dutyholders to comply with their duties and can be considered as the equivalent of manufacturers' technical literature and data sheets.

1.5 Absolute, practicable and reasonably practicable explained

Health and Safety law recognizes two categories of dutyholder:

- employers
- employees

Both are given three 'standards' or 'levels' of duty which are:

- absolute
- practicable
- reasonably practicable.

Absolute

An absolute duty is one that must be done. It is as simple as that.

In practice, absolute duties are rare but can arise where risk of injury or death cannot be avoided unless safety precautions are implemented. It is hard to imagine circumstances where this might arise, and if they did then the employer would be duty bound to implement appropriate safety precautions regardless of cost. An absolute duty is recognizable from the terminology used.

Thus, Regulation 3 of the Management Regulations says:

(1) Every employer shall make a suitable and sufficient assessment of –

(a) the risk to the health and safety of his employees to which they are exposed whilst they are at work; and
(b) the risks to the health and safety of persons not in his employment arising out of or in connection with the conduct of him by his undertaking

The use of the word 'shall' means what it says and places an absolute duty on all employers.

Practicable

A practicable duty is one which takes account of the latest technology, knowledge and equipment if it is possible to apply it (*Schwalb* v *Fass* (H & Son) [1946 175LT345]). Cost, difficulty and convenience are factors but the measure is, that if it can be done then it must be done. It is therefore less strict than absolute.

Practicable duties are again recognizable from the terminology, which usually takes the form: 'every employer shall, so far as is practicable, . . .'

Reasonably practicable

The original parameters for reasonably practicable were set in *Edwards* v *The National Coalboard* [1949] 1All ER 743 where the Judge said:

Reasonably practicable is a narrower term than physically possible, and seems to me to imply that a computation must be made by the owner in which the quantum of risk is placed on the scale and the sacrifice involved in averting the risk, whether in money, time or trouble, is placed on the other . . . the questions the owner has to answer are . . . what measures are necessary to prevent breach? Are these measures reasonably practicable?

A reasonably practicable duty is usually phrased: 'every employer shall, so far as is reasonably practicable, carry out . . .'

Most health and safety duties come into the category of reasonably practicable. A common means of illustrating the notion is to show a set of scales with cost on one side and benefit on the other. If cost heavily out weighs benefit then it is not reasonably practicable to implement the proposed measure. Unfortunately, this offers little practical guidance to many employers because the more likely scenario is that the scales will not be heavily out of balance.

The duties can be wide ranging. The case of *Associated Dairies* v *Hartley* [1979] IRLR 171 provides an illustration. The company had 66 depots around the UK and had calculated that if it provided safety boots to all of its employees who might need them, the cost would be £20 000 in the first year and £10 000 thereafter. So the company introduced a scheme whereby as a result of a Factory Inspector's improvement notice, it offered its employees a credit facility to buy safety footwear for £1 per week each (the improvement notice had stipulated free footwear). The company had decided after a cost/benefit analysis that it was not reasonably practicable to provide the footwear free of charge.

An accident happened at one of the company's premises when a roller truck injured an employee's toe. He had not purchased and was not wearing safety footwear. Only six of the staff at the premises concerned had taken up the firm's offer. The outcome of the case was that the firm argued that it had acted in accordance with trade practice and had done enough to demonstrate reasonably practicable. The Judge agreed with them.

1.6 HSC, HSE explained

The most important influence on present day health and safety law was the Robens committee (1970–72) and in particular its report entitled 'Safety and Health at Work', which led to HASWA 1974.

In turn, the Act established the current structure whereby the **Health and Safety Commission** has overall responsibility for securing the health, safety and welfare of people at work, and members of the public affected by those at work. It is advised and assisted by the **Health and Safety Executive**, which also has statutory responsibility for enforcing health and safety law. In order to perform its functions HSE employs about 4300 people including inspectors, scientists, technologists, medical experts and policy advisers. HSE reports to the government, and to HSC.

Consequently, few dutyholders will come into contact with HSC, but all may deal directly with HSE. It is not an analogy that finds favour with the Executive, but it is the 'policeman' of health and safety law.

Its ethos is one of encouragement, courtesy and co-operation, and it is one which it practices as well as it preaches. If HSE can see that a dutyholder is trying to comply with the law, even though he may have got it wrong, they will usually give the benefit of any doubt. Only in cases of flagrant non-compliance do they seek to use their full powers, which can be draconian.

A case the writer remembers well involved an HSE inspector who wished to visit a building site within a large RAF base in the south of England. He was denied access by the military police at the gate. He explained who he was and identified himself. The officer of the watch, acting on the base commander's instructions, continued to deny him access on the grounds that the site he wished to visit was subject to the Official Secrets Act. Within minutes of the second denial, the base commander was instructed by telephone from sources within the Ministry of Defence that he was acting unlawfully and was exposing himself to immediate arrest and the probability of criminal charges for breach of the Health and Safety at Work etc. Act 1974.

Suitably chastened, the base commander facilitated access and the HSE inspector then found undisclosed asbestos on the site in question, which was a large refurbishment project. The discovery was a very serious matter which resulted in the immediate cessation of work on the site, and although in reality it had nothing to do with the base commander, he had a very hard time in subsequently explaining his actions in denying access to the inspector.

In any dealings with HSE, it is important to remember that they enforce the criminal law, not civil law. For that reason, it is rarely a good idea, if subject to an HSE investigation, to wave a contract at an inspector whilst attempting to defend oneself along the lines of 'we did it this way because it told us to do so in the contract'.

Very often, contracts are written by people with no experience, or even worse, a little experience, of health and safety law. This axiom extends to many lawyers; they may be very good at civil law but their knowledge of the criminal code and the HASWA/M HASWA framework can leave much to be desired.

Consequently, they tend to write clauses into contracts which can:

- require one of the parties to do less than is required of them under statute;
- oppose the requirements of the criminal code;
- refer to irrelevant laws;
- refer to repealed laws, because the contract has been 'word processed' from a previous job.

It is a basic tenet of health and safety law that all dutyholders are responsible for their own actions.

Very often, in waving a contract at an inspector, all one is doing is providing him with clear, written evidence of the way in which the law has been broken. It is not a defence to argue ignorance of the law, or that one's actions as a dutyholder were dictated by others. The Derby City Council case gives a clear example.

The Health and Safety Executive

The HSE is specifically the enforcing authority in respect of the following activities:

1 Any activity in a mine or quarry.
2 Fairground activity.
3 Any activity in premises occupied by a radio, television or film undertaking where broadcasting, recording, filming or video recording is carried out:
4 (a) construction work;
   (b) installation, maintenance or repair of gas systems or work in connection with a gas fitting;
   (c) installation, maintenance or repair of electricity systems;
   (d) work with ionizing radiations.
5 Use of ionizing radiations for medical exposure.
6 Any activity in radiography premises where work with ionizing radiations is carried out.
7 Agricultural activities including agricultural shows.
8 Any activity on board a sea-going ship.
9 Ski slope, ski lift, ski tow or cable car activities.
10 Fish, maggot and game breeding.

The HSE is the enforcing authority against the following and for any premises they occupy including parts of the premises occupied by others providing services for them:

11 Local authorities.
12 Parish councils and community councils in England and Wales.
13 Police authorities.
14 Fire authorities.
15 International headquarters and defence organizations and visiting forces.
16 United Kingdom Atomic Energy Authority.
17 The Crown.
18 Indoor sports activities.
19 Enforcement of the Health and Safety at Work activity section 6 (manufacturers duties).
20 Off-shore installations.

Where the Health and Safety Executive is the enforcing authority, the hierarchy of enforcement is as follows:

Improvement notices

An inspector may serve an improvement notice if he is of the opinion that a person is:

(a) contravening one or more of the relevant statutory provisions, or
(b) has contravened one or more of those provisions in circumstances that make it likely that the contravention will continue to be repeated.

In the improvement notice the inspector must:

1 state that he is of the opinion in (a) and (b), and
2 specify the provisions in his opinion contravened, and
3 give the particulars of the reasons for his opinion, and
4 specify a period of time within which the person is required to remedy the contravention.

The person on whom the notice has been served then has 21 days to comply or to lodge an appeal with an industrial tribunal. At the end of the 21 days the inspector will return if no appeal has been lodged and if the notice has not been complied with it can have serious penal consequences. The maximum fine which can currently be imposed on summary conviction is £5 000 (Criminal Justice Act 1991). In certain circumstances, however, magistrates can impose fines of £20 000.

Prohibition notices

If an inspector is of the opinion that workplace activity involves or will involve a risk of serious personal injury, he will serve a prohibition notice on that person. A prohibition notice is different from an improvement notice in two important ways:

1 It is not necessary for an inspector to believe that a provision of the Health and Safety at Work Act or any other statutory provision is being, or has been contravened.
2 It is enough for an inspector to anticipate danger.

In a prohibition notice an inspector must:

(a) State that he is of the opinion there is a hazardous activity or state of affairs.
(b) Specify the matters which create the risk.

(c) Where there is an actual or anticipated breach of any regulations, state that he is of the opinion that this is so and set out the reasons.

(d) Direct that the activities referred to in the notice must not be carried out, on, by or under the control of the person on whom the notice is served.

An improvement notice gives the person on whom it is served time to correct the defect or offending situation. A prohibition notice, which is a direction to stop the work or activity in question rather than to put it right, can take effect immediately on issue.

Accordingly, it is a question of either an improvement notice or a prohibition notice. If an offence is thought to have created serious risks to life or disclose flagrant disregard of health and safety duties or alternatively the offence was persistent, the matter can be laid before a Crown Court which can impose an unlimited fine and jail sentences.

Local authorities

In some cases, local authorities are the enforcing authority for certain activities and the main ones are scheduled below:

1 Sale or storage of goods for retail, wholesale, distribution including motor car tyres, exhausts, windscreens or sunroofs.
2 Display or demonstration of goods at an exhibition being offered or advertised for sale.
3 Office activity.
4 Catering services.
5 Caravan or camping sites and similar temporary residential accommodation.
6 Consumer services provided in a shop.
7 Coin operated units in laundrettes.
8 Baths, saunas, solariums, massage parlours, premises for hair transplants, skin piercing, manicuring or other cosmetic services and therapeutic treatments.
9 Practice or presentation of arts, sports, games or other cultural recreational activities unless carried out in a museum, art gallery, theatre or caves.
10 Hiring out of pleasure craft for use on inland waters.
11 Undertaking.
12 Church worship or religious meetings.
13 Temporary grandstands.
14 Care, treatment, accommodation of animals, birds or other creatures with certain exceptions.

1.7 Health and safety legislation enforcement and penalties

As seen above, the powers of the Health and Safety Executive are broad and allow inspectors to:

- enter any premises where work is carried on without giving notice;
- talk to people at work;
- take photographs and samples;
- impound damgerous equipment and substances;
- issue advice or warnings;
- issue improvement notices;
- issue prohibition notices;
- mount prosecutions against individuals, corporate bodies, nationalized industries or local authorities;
- investigate accidents;
- seize evidence;
- consult and appoint experts.

On average about 25 000 notices are issued and 2600 prosecutions are made each year, 90 per cent of which are successful.

1.8 Chapter summary and conclusions

Health and Safety Law in the UK has developed over two centuries and is something of a patchwork quilt, evolving against the background of the Industrial Revolution.

Laws are made to reduce the cost of accidents in terms of human suffering, time, money, inconvenience and general business interruption. Nobody really knows the cost of accidents in these terms. It is put at around 2 per cent of gross national product in monetary value and direct measurable costs, but case studies show that it is much higher when human factors and indirect costs are taken into account. This latter approach is known as loss control.

Laws are made in the criminal code and the hierarchy of the law starts with Acts, which are made by parliament and are the law of the land. These set out the requirements of the law in a general way and require a set of regulations, which can be made by ministers, to set out the minimum standards to be achieved by dutyholders.

The law places duties on all employers and all employees in all workplaces and extends to members of the public affected by those at work. These duties are structured in to three levels: absolute, practicable and reasonably practicable, each of which has been defined by the Courts.

The intent of the legal framework is to achieve similar standards of health and safety management in all workplaces and in all workplace activity.

Approved codes of practice are sometimes used to give advice to dutyholders on how to achieve the minimum standards set out in Regulations, and whilst they are not law, they have the force of law.

Guidance notes, of which there are about 1500, have no legal standing and are also intended to give practical advice on how to comply with the law.

The Health and Safety Executive enforce the law and have far reaching powers, which they seek to use in a positive way by encouraging dutyholders to comply.

Breach of the law is a criminal offence and penalties can include fines and jail sentences. A successful criminal action will almost certainly lead to a successful civil action for damages against persons convicted of breach of health and safety law.

The differences between the criminal and civil codes are not widely understood. Most people are more familiar with the latter than the former, and frequently fail to understand that responsibility for a statutory duty rests where it falls. It cannot be passed to someone else in the same way as, for example, a business responsibility, more usually described as a risk.

Contracts are made under the civil code and a contractual obligation must not oppose a statutory duty. To that extent, there is a clear danger in dealing with health and safety in contract.

2

The overall architecture of health and safety law

2.1 Introduction

In the United Kingdom the overall architecture of health and safety law follows the principles set out by the Robens Committee (1970–72) and established in the Health and Safety at Work etc. Act 1974. It is briefly described in the following model:

- **General duties:** The creator of risk controls it.
- **Goal setting regulations:** Establish standards.
- **Codes of practice:** Set out how standards can be achieved through good practice.
- **HSE guidance:** Give advice on how to comply.

This is recognized as a simple practical system which has widespread support across business and industry.

Unfortunately, as briefly described in Chapter 1, it is complicated by two significant factors which are looked at in more detail here. First, it sits alongside the old, pre-1974 domestic legislative framework comprising mainly prescriptive laws and regulations. Second, since 1974, the UK has entered the European Union, which has subsequently become the main source of legislative change in recent years.

2.2 Prescriptive law making 1800–1974

Prescriptive laws tell dutyholders what to do, and in many cases how to do it. They evolved against the background of the vast social, political and economic changes which occurred in the UK between 1800 and 1974.

It is worth remembering that in this period we changed from a largely agrarian society with a population of around 13 million to a highly industrialized society with a population of around 55 million. New industries and new technologies came and went. Some lasted, but changed their methods beyond recognition. Thus, as an industry or business sector developed, so laws were made to regulate it, usually some considerable time after the initial development of the industry or technology took place.

Some lingering examples of old laws remain. The first Factories Acts were made in 1833 and were updated at regular intervals until 1961. An Act relating to shops was first made in 1866. The India Rubber Regulations of 1920 and Vitreous Enamelling Regulations of 1908 refer to technologies which once employed thousands of people, which now employ hardly anybody.

So, by the early 1970s the health and safety system resembled a vast patchwork quilt. Across business and industry, employers were required to provide varying health and safety management standards in different workplaces. Clearly, this was confusing enough and was further complicated by the fact that many employees in large sectors of business and industry were not covered by any health and safety legislation at all. For example, schools and educational establishments were outside the system.

The regulator's approach exacerbated the problem because they tended to modernize the law by means of a series of partial revocations and repeals which simply added to the chaos.

Some industries or technologies will probably always need prescriptive laws. For example, all mines must have two exits. If this was not a legal requirement, there would surely be a strong temptation to omit the second exit on cost grounds, an action which most people would find unacceptable.

On the other hand, the majority of industries and technologies require moderate health and safety management standards because there is nothing intrinsically dangerous or unsafe about the activities involved in them. For example, the average building society office is a fairly safe place, provided people work safely and use equipment as trained; so is a vehicle repair workshop or a department store or a modern car manufacturing plant and so on.

Nevertheless, the system as it was in the early 1970s was clearly in urgent need of reform, and in particular, the old prescriptive approach required a complete re-appraisal.

2.3 Goal-setting law making 1974 to the present

Reform arrived in the shape of Lord Robens and the Robens Committee (Safety and Health at Work, 1970–72). In 1974 there were about 50 Acts and 500 sets of regulations in place, all taking the old prescriptive approach. Robens' broad intention, expressed in the committee's report, was to sweep away all of this regulation and replace it with one Act, and new improved regulations, approved codes of practice and guidance notes. Underlying this, a new philosophy was to be used to underpin the approach, that of risk assessment, to be undertaken by those who own, manage or work in commercial and industrial undertakings. This extends to the self employed.

In this way, all workplaces, all employers, all employees and certain members of the public would be brought within the framework of the law. The real benefits arose from the standardization of approach. Now employers had to achieve the same standards of health and safety management in all workplaces, and all employees had to be treated the same. This still left room for a prescriptive approach in work activities that were unavoidably less safe than others, for example coal mining, but for the broad sweep of business and industry, the new approach was far better than its predecessor, and it was well received and warmly welcomed.

However, it was not possible to implement the new approach overnight and it was recognised that the two systems would co-exist for as long as it took to achieve the intention of the Robens Committee, and that this would take years. Unfortunately, this lead to further difficulty.

2.4 Volume, diversity and difficulty of legislation

In December 1992, Michael Forsyth, minister of state with responsibility for health and safety law at the time, invited HSC to review workplace health and safety legislation, and to advise the Government on:

- whether it was still relevant;
- whether it was all necessary in its existing form;
- whether it placed unnecessary burdens on business.

In May 1994, in response to the minister's invitation, HSC published its 'Review of Health and Safety Regulation: Main Report'. The fundamental question examined by the review was whether the costs generated by complying with the law are justified by the benefits.

In summary, the review explained that the costs are justified by the benefits, in some cases many times over, particularly when the 'loss control' approach outlined in Chapter 1 is taken into account.

However, the review identified some serious impediments to a full understanding of the health and safety system, and these can be considered under the following headings:

Volume

By 1994 the number of Acts had fallen to 28, and the number of sets of regulations had reduced to 367. However, some 56 approved codes of practice and 1500 guidance notes had been introduced, and the review commented that it had found strong evidence, particularly in small firms, that volume:

- requires time to hunt down legal requirements, and
- diverts management resources from practical issues.

Diversity

In practice, by 1994, only two laws applied to all work situations all of the time, namely:

- Health and Safety at Work etc. Act 1974, and
- Management of Health and Safety at Work Regulations 1992.

The application of the rest of the laws, regulations and approved codes of practice then depends upon the specific workplace activity. Therein lies the first problem for employers, that is, understanding which laws apply. There is no easy answer to this problem, it is simply a function of the competence of employers and the steps that they take to discover this information.

The second problem is to be found in the existence, side by side, of both old-style prescriptive laws and new, modern goal-setting regulations. Again, there is little to be done other than to understand which regulations apply within the HASWA/MHASWA framework, and to implement them.

Difficulty

Difficulty, so HSC explained in their review, arose from a combination of volume and diversity, coupled with a lack of understanding of the law. Volume and diversity has been discussed above, but the lack of understanding of the law flows from the following causes:

- the hierarchy of the law, Acts, regulations and approved codes of practice is not widely understood;
- the levels of duty, absolute, practicable and reasonably practicable are also not widely understood.

For example, the term 'so far as is reasonably practicable' is in *fact*, a legal standard laid down by the decision in *Edwards* v *The National Coalboard* [1949] 1All ER 743. Not many people understand that it is a legal standard, and it creates uncertainty in the mind of the average dutyholdcr.

HSC recognize in the review that many employers simply prefer to be told what to do. They say that the real problem lies, not so much in understanding the terminology, as in understanding the risk assessment/cost benefit analysis approach.

However, the point is probably academic, for it amounts to the same thing. Dutyholders have difficulty in complying with the law because there is a lot of it, they do not understand the way in which it is written, or the terminology used. They feel that it is time-consuming trying to understand the legal requirements and to complete the paperwork generated by the law. Again, at the time of the review in May 1994, Table 2.1 summarizes the fact that seven Acts and 183 sets of regulations created a requirement for 340 separate pieces of paper. Most of this is perceived to be unnecessary, or at the very least, a burden.

HSC's response is that 'No single employer is actually required to fill in 340 pieces of paper. In fact, even the most complex workplace activities require only a fraction of that paperwork.'

The pressure for form filling does not originate solely with the regulatory authorities. A high proportion arises from other sources, which can include:

- misunderstanding of legal requirements;
- internal business purposes (personnel, inventory etc.);
- insurers;
- quality standards;

Table 2.1   Paperwork required by law

Type of paperwork	Regulatory	Recommended
Notification	38	0
Record	91	11
Appointment	11	32
Written application	9	0
Information	5	56
Licences	9	0
Display of abstracts	78	0
Total	**241**	**99**

- advice from over zealous consultants;
- contract selection and appointment procedures.

The main concerns, when examined and analysed by HSC, in fact stem from ten sets of modern regulations including COSHH, RIDDOR, MHASWA and the rest of the 'six pack' which are not well or widely understood. These are all dealt with in a later chapter.

The examination and analysis reveals that many companies go 'over the top' when dealing with these regulations, either because their systems are excessively paper based, or because they do several risk assessments where one would suffice, that is, one for each set of regulations instead of a consolidated one for a single workplace activity.

'Newness' can be a factor as it takes time for employers to understand and integrate new laws into their organizations.

2.5 Future developments: HSC review, May 1994

In summary, the review of 1994 revealed that there is widespread support for the maintenance of the existing health and safety system in the UK. However, HASWA provides that legislation passed before 1974 should be: 'progressively replaced by a system of regulations and approved codes of practice'.

In time, there will be one Act and perhaps more than a hundred sets of regulations accompanied by approved codes of practice and guidance notes. However, the eventual number of laws, and their content will be heavily influenced by developments within the European Union, which the UK entered in 1979.

2.6 The European dimension

There are currently 13 member states in the European Union and it is reasonable to assume that the ranks will be swelled in due course by new members from the former Eastern Bloc and even Russia itself.

Britain and continental Europe have differing traditions of legal interpretation. A fundamental tenet of English Law is 'words mean what they say'. Thus, an English court would not see the need to interpret the law which said 'no person shall drink and drive'.

Continental courts, on the other hand, have always shown willingness to take circumstances into account. Hence, for example, French understanding of 'crimes of passion'. If a Frenchman were to murder his wife for no good reason, a French court would almost certainly find him guilty and would probably give him a maximum sentence, but if he were to murder his wife because he had found she was having an affair, the court might take the view that maybe he was not so guilty and that he deserved a much shorter sentence. Such matters depend upon the judge, to an extent that would not be encountered in the UK. This is not because the judges are any better or worse than each other, it is because the legal traditions are different.

Many continental countries are therefore comfortable with prescriptive laws, for they know that in practice, laws and absolute requirements may be moderated by the decisions of the courts, an approach diametrically opposite to that of the UK.

Legal frameworks vary considerably from member state to member state. The source of law in Spain is its constitution of 1978 and it operates seven court systems. Germany's law is based on its constitution of the 21 May 1949 and it has a court system comprising five branches. The source of law in Belgium is its constitution adopted in 1831 and based on the Napoleonic Code. It has a court system comprising six different branches whilst on the other hand France has a constitution which was adopted in 1958 and a court system comprising two jurisdictions, judicial and administrative. Denmark still uses Nordic law and the National Code of 1683 as its source of law, whilst The Netherlands, like the UK, relies on custom as well as laws and treaties.

However, all member states share, besides membership of the European Union, two other common factors:

- they all recognize a criminal code and a civil code;
- they all implement EU law into their domestic legislative framework in similar fashion.

The Common Market was originally founded to implement free trade in coal and steel between its six member states, comprising Germany, France, The Netherlands, Italy, Luxembourg and Belgium. Its origin was the Treaty of Rome 1957. Subsequently, it has expanded and evolved in to the European Union which has much broader aims and objectives; some would say the 'Federal state' of Europe.

The EU has four main legislative institutions. They are:

The European Parliament

This is an elected body which draws its members from its member states. At present, it can only make laws if it has consulted with the Council of Ministers and its role is consultative and advisory.

Council of Ministers

This is the final decision-making body and effectively is the government of the EU. It considers proposals from the European Commission after which it must consult with Parliament and is therefore a political body.

European Union Commission

This consists of 17 independent members with the five larger member states each appointing two commissioners and all of the other members one each. Its purpose is to propose EU law to the Council of Ministers and as such, it is not political but is the executive arm of the EU. Its proposals normally take the form of directives.

European Court of Justice

The court sits in Luxembourg and comprises 13 judges and six advocates general. It administers community law by referral (usually written rather than oral) sent to it by the national courts of member states. In keeping with continental tradition, it tends to look at the intent of the law, rather than the letter of it, and it is said that all of its decisions are free of dissenting opinion.

It has no jurisdiction in the UK in respect of the Social Chapter of the 1991 Maastricht Treaty because the UK Government, under the then Prime Minister John Major, did not sign the chapter. However, this does not apply to health and safety legislation because it is subject to qualified majority voting (QMV), and we accepted this part of Maastricht anyway.

European Union Law

In the case of any conflict between EU law and any national law, the former prevails. Most member states have some form of internal legal mechanism to act as 'watchdog' over EU law and the manner in which it impacts on national law. In the UK the Houses of Commons and Lords have a joint committee set up by Parliament to undertake this task.

The hierarchy of EU law is:

■ treaties and Acts;
■ regulations;
■ directives.

Treaties and Acts

These include:

- The Treaty of Rome 1957;
- The Single European Act 1986;
- The Maastricht Treaty 1991.

Treaties and Acts in European law set out in a very general way the requirements of the law, which then require an internal statute detailing the expressed rights to be conferred by means of domestic legislation.

Regulations

Regulations become part of domestic law automatically and they have practically no role to play in health and safety.

Directives

Under Article 189 of the Treaty of Rome, directives are binding in all member states but members are free to decide upon the means and manner of implementing them.

Normally member states have two years in which to implement them although this can be varied. Failure to implement any directive can result in 'infraction proceedings' automatically set in hand at the European Court of Justice and severe penalties can be imposed upon non-compliant member states.

As far as health and safety is concerned, directives are the most important source of legislation in the UK. Implementation here can be by Act of Parliament but is usually by statutory instrument or regulation. This is because the regulation is nearly always issued under the Health and Safety at Work etc. Act 1974 which, as has been pointed out before, is intended to be the sole Act of Parliament on the statute books.

Therefore, domestic Acts implementing EU directives tend to be specific; for example, the Product Liability Directive was implemented in the UK by means of the Consumer Protection Act 1987.

There are two types of directive:

- **Framework directives:** set out a general framework of controls and operate in a similar fashion to a domestic act.
- **Daughter (satellite) directives:** set out specific features and operate in similar fashion to domestic regulations.

Members must not only implement directives, they must also comply with them.

European Union objectives

From the outset, Article 100 of the Treaty of Rome laid down a strategy within the internal market of achieving 'approximation' and 'harmonization' of laws and practices.

In 1986, Articles 8a, 8b and 8c of the Single European Act called for the progressive dismantling of social barriers, and Article 118a required improvement in the working environment with regard to health and safety of workers.

Thus, legislation is now in place both in the EU and all its member states which compels them to achieve 'approximation' and 'harmonization' as originally envisaged in the Treaty of Rome but extended by subsequent treaties and Acts and by qualified majority voting.

In health and safety terms, this means that:

- all employers are duty bound to offer similar workplace standards across the EU;
- all employees are to be treated the same;
- all employees have similar rights and duties.

Thus, duties imposed in the UK by, say, the Health and Safety at Work Act 1974 and the Management of Health and Safety at Work Regulations 1992 apply directly in the UK and by extension to all EU members who must implement their own domestic version of these regulations.

In June 1989, the Council of Ministers adopted a framework directive on its general approach to safety and health of workers at work, and in November 1989 it adopted a series of related directives on specific aspects of safety and health. These are minimum standards. Member states are free to exceed them and the directives include:

Safety and health of workers at work – the framework directive (89/391/EEC)

Adopted June 1989; required to be adopted in member states: 31 December 1992

This is the directive that lead to the implementation of the 'six-pack' in the UK and its domestic equivalent across Europe. The framework directive

(89/391/EEC) can therefore be seen to have been one of the most important legislative developments emanating from the EU and impacting on health and safety law in the UK in recent times.

Like all directives, it is notable for the manner in which it is written. It is simple. It does not use jargon or unnecessary technical terms and it does not refer to civil law arrangements such as forms or types of contract or procurement methods. It simply looks at workplace circumstances as they might occur in all countries regardless of the civil or common law arrangements which might exist in any one country and it sets out, in a very general way, the objectives that have to be achieved. It then leaves member states to achieve compliance in a manner that fits their own legal systems and procedures. Sometimes, directives can be accompanied by annexes which are used to set the minimum standards which must be achieved where the legislators believe that a little more precision is required.

The directive contains 18 articles of which articles 6 and 7 are key. The bullet points of these articles are set out below:

- Employers are responsible for safety and health of workers at work.
- Employers must designate a worker or outside service to deal with prevention and protection of occupational risks.
- Worker consultation is compulsory.
- Employers must bear the costs of safety and health measures.
- Workers must be given adequate time and not disadvantaged.

The European dimension: summary

According to the European Commission, there are more than 4.5 million accidents per year in the EU, of which 8000 are fatal. The total direct insured costs to national social security systems is believed to be around £14 bn and readers are reminded that this does not include uninsured indirect costs where there are no injuries or deaths.

The object of the Workers Framework Directive 89/391/EEC is 'to introduce measures to encourage improvements in the safety and health of workers at work'.

It goes on to say that the directive:

'contains general principles concerning the prevention of occupational risks, the protection of safety and health, the elimination of risk and accident factors, the informing, consultation, balanced participation in accordance with national laws and/or practices and training of workers and their representatives as well as guidelines for the implementation of the said principles.

The effect of this directive is to introduce a community-wide risk assessment procedure to all work-related activities, and to implement procedures for dealing with the identified risks in the safest possible manner.

A shrewd observer might immediately notice that this seems to be a complete about-face on behalf of continental Europe which has moved away from its old philosophy of 'every worker is responsible for their own safety' to an approach that is much closer to that which has been in place in the UK since 1974.

It is instructive to note that the directive drafting body considers that unsatisfactory management plays a major role in occupational health and safety across the EU, and the directives are drafted in such a way as to address the causes. In so doing, they recognize that the causes include:

- poor record keeping;
- widespread use of subcontracting;
- poor information dissemination;
- inadequate training;
- inadequate allocation of resources including time;
- lack of monitoring and review procedures;
- inadequate co-ordination;
- inadequate co-operation;
- unequal and unequitable risk apportionment;
- lack of competence;
- unsatisfactory approach and attitudes towards health, safety and health issues.

2.7 Chapter summary and conclusions

In the UK the overall architecture of the law comprises Acts, regulations, approved codes of practice and guidance notes, which enshrine a goal-setting/risk-assessment philosophy which has been in place since 1974.

This sits alongside the old prescriptive approach to law making, which tells dutyholders what to do, almost an opposing approach. This is confusing for dutyholders, who must learn to deal with both approaches for as long as they co-exist. In time, only the goal-setting/risk-assessment approach will be used, most other old or prescriptive laws will be swept away unless they are absolutely necessary.

The volume of law presents difficulty to dutyholders, as HSC identified in their review of 1994, and the terminology used in terms of absolute, practicable and reasonably practicable requirements exacerbates the difficulty because many dutyholders do not understand the legal significance of the terms.

Paperwork can be a burden. In total health and safety laws can give rise to 340 separate pieces of paper, but HSE point out that this situation will never arise in any single workplace, and in any event, in their view, non-regulatory sources place a bigger paperwork burden on employers. Nevertheless, HSC recognize that some concerns are justified and in response they are proceeding to reduce the number of laws on the statute books, standardize the approach, explain the law better and reduce any paperwork that they create.

HSC also recognize the impact of EU law in the UK and point out that it will play an increasingly important role. Membership of the EU requires all member states to achieve approximation and harmonization. In practice, this means that, despite their different legal systems, each member state much achieve similar workplace conditions and similar health and safety management standards in a similar timescale, or risk severe penalties which can be imposed by the European Court of Justice.

As a result of the Workers' Framework Directive 89/391/EEC, approximation and harmonization are now a reality, and there is now in place a community-wide risk assessment philosophy similar to that enshrined in HASWA/MHASWA.

3

HASWA/MHASWA framework

3.1 Common law duties

In simple terms, the Health and Safety at Work etc. Act 1974 is a formal statement of common law duties of care. We owe each other certain duties of 'reasonable care' and if we neglect those duties, then we may be guilty of 'negligence'.

Thus, even if HASWA had not been made, the common law duties that would (and do) exist are:

Employers must take reasonable care to:

- avoid injuries to employees;
- avoid diseases to employees;
- avoid deaths of employees;
- provide a safe place of work;
- provide safe access and egress to the workplace;
- provide safe plant and equipment at the workplace;
- maintain safe plant and equipment at the workplace;
- provide safe work systems;
- operate safe work systems;
- provide 'competent person' assistance where necessary;
- safeguard the safety, health and welfare of members of the public affected by work activity.

Employees have a general duty of care:

■ to themselves;
■ to their colleagues at work;
■ to members of the public affected by their work;
■ to their employer.

Until relatively recent times, the notion of '*volenti non fit injuria*' (to one who is willing no harm is done) prevailed, and led to the general view, which was also held by the judiciary, that if a worker was injured at work it was mainly his own fault; he did not have to be there, but if he chose to be, then he accepted the risks that went with it.

This was re-enforced by the doctrine of *common employment* which held that employers could not be held liable for injuries caused by a fellow employee at work. However, from around the turn of the twentieth century, these notions and doctrines came under increasing challenge and, by the early part of the twentieth century, were totally discredited.

3.2 Duties under HASWA 1974

A casual comparison of the common-law duties with an abbreviated version of the Health and Safety at Work Act reveals many similarities.

First, the aims and objectives of the Act are:

■ to involve everyone in health and safety;
■ to achieve the health, safety and welfare of persons at work;
■ to protect the health and safety of all people who might be at risk from the activities of persons at work;
■ to place the burden of legislation compliance firmly upon employers, but also upon their employees to ensure that they comply.

The Act applies to:

■ employers
■ employees
■ some members of the public
■ the self employed
■ designers
■ importers, and
■ suppliers.

The Act is organized in the following way. It comprises four parts:

- **Part 1:** Deals with work place health, safety and welfare.
- **Part 2:** Deals with the Employment Medical Advisory Service.
- **Part 3:** Amends the laws relating to the Building Regulations.
- **Part 4:** General and miscellaneous provisions.

The main provisions of the Act are as follows:

Employers

- To ensure the health, safety and welfare of all employees.
- To provide a safe place of work.
- To provide safe access and egress to the place of work.
- To provide a safe work environment.
- To provide adequate welfare facilities.
- To provide safe plant and equipment for use at work.
- To maintain safe plant and equipment for use at work.
- To provide safe work systems.
- To maintain safe work systems.
- To ensure there is no risk in material or substance handling.
- To ensure that risk in material or substance handling is eliminated, reduced or controlled.
- To provide information on work activity risks to employees.
- To provide instruction and training to employees.
- To provide adequate supervision.
- Where there are five or more employees in a workplace, provide a written safety policy.
- To provide a written report on health and safety activities in the shareholders' report.

Self-employed Persons' duties

Self-employed persons can have the duties of both employers and employees; much depends upon the way in which the self-employed person is employed.

A general duty is placed on all self-employed people to conduct their work in such a way as to ensure that they and other persons are not exposed to risks to health and safety.

Employees

- To take reasonable care of their own health and safety.
- Take reasonable care of the health and safety of others who may be affected by their acts.
- Co-operate with the employer.
- Do not mis-use or mistreat work equipment.

Manufacturers, designers, importers and suppliers

A general duty is placed on manufacturers, designers, importers and suppliers in respect of any article, materials or substances for use at work. The duty is to ensure, so far as is reasonably practicable, that articles and substances are safe and without risks to health when being used, set, cleaned or maintained by persons at work.

A further duty exists to ensure that arrangements are made to carry out any necessary testing, examination and research, and the duty extends to providing adequate information about any conditions needed to ensure safety in use.

A general duty is placed on installers or erectors of any article for use at work to ensure, so far as is reasonably practicable, that the article is safe and without risk to health when used by persons at work.

Noise level information must be given in accordance with the requirements of the Noise at Work Regulations.

Written policy statements

Employers with five or more persons are under a statutory duty to prepare and keep updated a written safety policy, and to communicate this to all employees.

The written safety policy should cover the following aspects as a minimum:

- policy
- organization
- arrangements
- updating
- communication.

Penalties: 1997

Penalties are, from time to time, updated and revised. At the time of writing (June 1999) the penalties listed in Table 3.1 were in force.

Table 3.1 Enforcement penalties

Scales	Penalties
Scale 1	A £20 000 fine at a magistrates' court or sheriffs' court or an unlimited fine at a crown court or a high court in Scotland.
Scale 2	A £5000 fine at a magistrates' court or a sheriffs' court or an unlimited fine at a crown court or a high court.
Scale 3	A £5000 fine at a magistrates' court or a sheriffs' court.
Scale 4	A £20 000 fine or six months imprisonment, or both, at a magistrates' court or sheriffs' court, or two years imprisonment, or an unlimited fine, or both, at a crown court or a high court.
Scale 5	A £5000 fine at a magistrates' court or a sheriffs' court, or an unlimited fine at a crown court or a high court, plus a crown court may impose two years' imprisonment or an unlimited fine or both.

Commentary

If casual comparison of the common law duties with part 1 of the Act is made, it will reveal many apparent similarities. However, there are a number of important differences.

First, the Act is criminal law and brings with it criminal penalties for breach. Second, common law duties extend to 'reasonable care', a standard of conduct which can be argued, and in any event, is somewhat less than 'reasonably practicable'. Third, most HASWA duties are 'reasonably practicable', a standard which is much more difficult to argue, and which can be tested against regulations, approved codes of practice, case law and precedent.

Fourth, the provisions of the Act are much more detailed than the common law duties, although the Act is written in a very general way. Fifth, the Act is supplemented by a number of regulations. Thirty-six sets were issued between 1974 and 1993, culminating in a further six – the 'six-pack' – on 1 January 1993. The 'six-pack' included the **Management of Health and Safety at Work Regulations 1992** and **Approved Code of Practice**, the regulations which give full practical affect to the Act.

New regulations continue to be issued. The effect of these, and pre-1974 regulations, is far reaching, covering nearly every conceivable workplace situation in the UK.

3.3 Duties under MHASWA 1992 and ACoP

These are, without doubt, the most important set of regulations currently on the statute books because they affect everyone at work, all of the time, and because they give effect to the Act. They were issued by the Secretary of State for Employment using his powers under Section 16(1) of the Act. They are accompanied by an Approved Code of Practice (ACoP) and came into force on 1 January 1993.

In total, there are 17 Regulations and one Schedule.

Regulation 1, as always, deals with **citation and commencement**, and **interpretation**, which deals with statutory definitions and explains what is meant by the terminology in the rest of the regulation.

Regulations 14–17 deal with procedural and administrative matters arising from the issue of the regulations.

Regulations 3–12 are, therefore, the 'working' parts of the law and are summarized below:

Regulation 3: Risk assessment
Regulation 4: Health and safety arrangements
Regulation 5: Health surveillance
Regulation 6: Health and safety assistance
Regulation 7: Procedures for serious and imminent danger
Regulation 8: Information for employees
Regulation 9: Co-operation and co-ordination
Regulation 10: Host employers/self-employed persons undertakings
Regulation 11: Capabilities and training
Regulation 12: Employees duties
Regulation 13: Temporary workers.

The regulations themselves, and the accompanying ACoP are goal setting, and written in a general way because they have to cover all workplaces. Setting aside the health and safety function, the regulations and ACoP together comprise good business practice. If one were to simply follow the contents of the Management Regulations in running any business, it would be a well run business which would also be safe and healthy.

Many of the features of good business practice are indistinguishable from good health and safety management. Indeed, some experienced people have argued that there is no such thing as 'health and safety' in its own right, it actually comes free as part of a much broader approach to efficient work practice.

3.4 Case study explaining aspect of MHASWA

It might be instructive to consider a brief case study, illustrating several of the major features of the HASWA/MHASWA framework.

Figure 3.1 shows a typical arrangement where the HASWA/MHASWA framework applies in the UK (as an aside, its domestic equivalent also applies in each member state throughout the EU).

A client, perhaps an automotive component manufacturer, has decided to have some building work done at his premises. By way of example, assume the existing premises are 100 000 sq. ft, made up of 20 000 sq. ft of offices and 80 000 sq. ft of manufacturing space.

The client decides to refurbish and modernize the whole plant. Some of the work he wishes to arrange and manage himself, using direct specialist contractors, other parts of the work he wishes to have done by a building contractor.

Figure 3.1 depicts typical contractual and organizational arrangements the various parties might enter into. The building works might be undertaken by a suitably experienced and competent building contractor, who would enter into an appropriate contract with the client.

The building contractor might then undertake the works using a combination of specialist subcontractors and suppliers (heating, electrical and ventilation engineers, etc.) and labour employed in various ways, either directly, or as

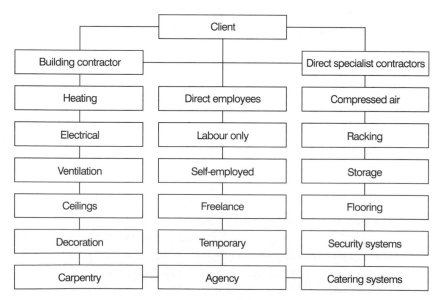

Figure 3.1 Organization chart for typical shared workplace arrangement

labour only or self-employed people etc., usually working on some sort of price arrangement, all bearing the cost of their own employment.

The direct works might be undertaken by the client using a series of specialists each on a separate contractual arrangement with the client, reporting directly to the client, but working with the other direct contractors, and probably the building contractor and his subcontractors and their labour force. Very often, all parties will share the same work places at the same time and will be required to carry out and co-ordinate their work around the client's continuing occupation and use of his premises.

A point to remember is that the specialist contractors may also use a combination of directly employed, labour-only, self-employed, freelance, temporary and agency staff to assist them in carrying out their work. Often it will be the same pool of labour – because it is local – being used by the client, the building contractor and his subcontractors.

Sometimes, the client may use a direct contractor to install, say, a compressed air system, and under a different contractual arrangement, the building contractor may use the same organization at the same time but in a different part of the premises to undertake, say, the heating subcontract.

Prior to the Management Regulations, there was no clear cut address of this entirely typical situation and it was obviously unsatisfactory because no single person or organization bore responsibility either in statute or under contract for managing and controlling the arrangements.

The Derby City Council case (Chapter 1) had established that health and safety could be the responsibility of more than one party, but the problem was that without a set of regulations to set objectives, and an ACoP to offer guidance, the parties to any arrangement had no means of testing them against an established legal standard.

In the case study arrangements and Figure 3.1, who is the party best able to manage health and safety whilst the refurbishment programme is in hand? Many people might say that it is the client, but it is axiomatic that the majority of clients would not agree with that, arguing that they do not have the skills necessary to undertake the building work – that is why they use building contractors and specialist direct contractors – therefore it follows that they do not feel able to co-ordinate and manage the health and safety aspects of the building works and they would require the main contractor to undertake that task.

With the introduction of the Management Regulations, there is still no prescriptive requirement for an arrangement such as this, but Regulation 9 and in particular Guidance Note 61, make it clear that 'employers and self employed persons present should agree such joint arrangements ... as are needed to meet the Regulations requirements'.

In other words, a goal-setting approach but offering clear guidance on what 'joint arrangements' mean:

- appointment of a health and safety co-ordinator;
- the obtaining of competent advice, on complex sites.

Clearly then the Management Regulations and Regulation 9 give a starting point from which a management structure can be designed and implemented. The obvious parallel in the broad business environment is the use of a project manager to co-ordinate the various contractual and operational arrangements shown in the model. Many clients would not consider proceeding without a project manager in an arrangement such as this, but would happily do so without a health and safety co-ordinator.

The starting point for any project manager would be the design and implementation of suitable work breakdown systems and organization breakdown structures, and all that the Management Regulations are envisaging is that similar standards of project management should be applied to the health and safety arrangements in any project.

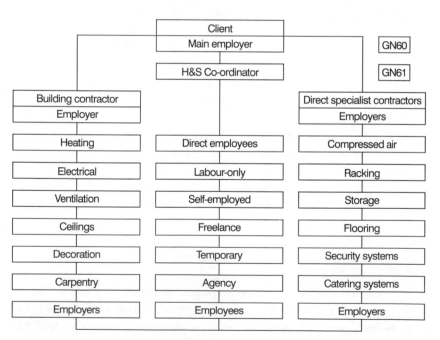

Figure 3.2 Organization chart illustrating the application of the MHASWA Regulations

So, having identified a 'main' or 'controlling' employer, that person or organization then takes on responsibility for the co-ordination of health and safety management arrangements on the project from start to finish. All other employers and employees and the self-employed have statutory duties to co-operate and comply with the reasonable arrangements the main employer makes. The role of main or controlling employer can be changed at any time and this would normally be done to fit in with the contractual arrangements that the parties wished to make.

In the model at Figure 3.1, terms such as client, building contractor, direct specialist contractors and the like have the meanings attributed to them under the civil law. In the criminal code, everyone in that model is either an employer or an employee, although one group could have the duties of both, that is labour-only and self-employed persons.

It is the main employer's responsibility in conjunction with these groups to clarify the way in which they are to be treated on a project by project basis.

If Regulation 9 and Guidance Note 61 are the starting point for the application of the Management Regulations in this particular arrangement, then the next step is Regulation 3, Risk Assessment.

3.5 Risk assessment regime overview

The hierarchy of risk control is as follows:

- **Hazard**: is defined as something with the potential to cause harm. This can include substances or machines, methods of work and other aspects of work organization.
- **Risk**: expresses the likelihood that the harm from a particular hazard is realized, and the extent of risk covers both workers and members of the public who might be affected by a risk and the possible consequences. Risk is therefore a measure of likelihood that harm will occur and its severity.

The duty of employers and the self-employed is to assess risks to workers and any others who may be affected by their undertaking. The methodology given at Guidance Note 5 in the Approved Code of Practice is:

- identify hazards arising in work activity;
- evaluate risk involved in the activity;
- assess the severity of the risk;
- decide measures to eliminate, reduce or control the risk;
- design and implement a safe work system;
- monitor, audit and review.

Risk assessments must be 'suitable and sufficient' and work activities involving five or more employees must be recorded in writing. There are a number of regulations currently in force which require risk assessment. Some are highly specialized and deal with issues such as major hazards, ionizing radiation, genetic manipulation and the like. These are outside the scope of this book.

However, there are also in force a number of regulations likely to affect most organizations, and these are:

■ Management of Health and Safety at Work Regulations 1992 (Management Regulations)
■ Manual Handling Operations Regulations 1992 (Manual Handling Regulations)
■ Personal Protective Equipment at Work Regulations 1992 (PPE)
■ Health and Safety (Display Screen Equipment) Regulations 1992 (Display Screen Regulations)
■ Noise at Work Regulations 1989 (Noise Regulations)
■ Control of Substances Hazardous to Health Regulations 1994 (COSHH)
■ Control of Asbestos at Work Regulations 1987 (Asbestos Regulations)
■ Control of Lead at Work Regulations 1980 (Lead Regulations)

All of these have a significant risk assessment component, but the Management Regulations are special in so far as they apply to all workplaces all of the time.

In addition to the above, the following regulations apply to construction:

■ Construction (Design and Management) Regulations 1994 (CDM)

Given that the Management Regulations apply to all workplaces, all employers, all employees and the self-employed, all of the time, all of the others apply if and when any workplace activity being carried out comes within the remit of individual regulations. Very often this is revealed by the risk assessment carried out under Management Regulation 3.

Hazard identification and risk assessment are looked at in detail in Chapter 5. The CDM Regulations are dealt with in further detail in Chapter 7.

3.6 Insurance issues arising from health and safety law

The detail of this matter is a highly complex area which is the province of lawyers. Nevertheless, employers must have sufficient understanding of the principles and the details to ensure compliance.

Overview

The principle set out in law is that, if people are injured or made ill as a result of work related activity, then they are entitled to be put back in the position they would have been in, to the extent that money can do so, had the event which lead to their injury or illness not occurred.

The means of achieving this is compensation, and there are two aspects to compensation, namely 'benefits' and 'damages'. The means of paying both is insurance.

Benefits

Benefits are payable under the social security system, which used to be known as National Insurance. The system can be thought of as a type of public insurance which is funded out of wage- or salary-related contributions made by employers, employees and tax payers.

It works on a 'no fault' basis, paying relevant benefits regardless of liability. A range of such benefits may be payable depending upon the nature, severity, length and effects of injury or illness. The rules are extremely complex and outside the scope of this book. The point made here is that the benefit system is meant to ensure that industrial accident/illness victims are not left destitute, and that they have some means of supporting themselves and their families whilst unable to work if injured or incapacitated.

There are many laws comprising the social security system, with the principal Acts being:

- Social Security Administration Act 1992
- Social Security Contributions and Benefits Act 1992
- Social Security (Industrial Injuries) (Prescribed Diseases) Regulations 1985

The system is not meant to put people back in the position they would have been had their accident or illness not happened, and in order to achieve that, they need to look elsewhere.

Damages

Many people have heard of 'torts'. Few people, other than lawyers, understand them and even lawyers struggle to explain them.

Torts deal with 'civil wrongs', and lawyers place them in categories, for example, the tort of nuisance or the tort of negligence. Damages may be

45

payable under a particular tort for a civil wrong if legal liability can be established. It is the liability that is important, and the damages (money) flow out of that.

In 1991 in the UK about 2.7 million businesses were registered. No one knows how many of them were actually trading, and around 1 million of them were not VAT registered, indicating that if they were trading, their turnover was less than £37 500. It is believed that about 90 per cent of active businesses employ fewer than five persons, either directly or indirectly.

If such a business were to be responsible for a serious accident resulting in injury or ill health to an employee, which led to long term injury or ill health, it is unlikely that such a business would have the resources to put that person back in the position he or she would have been in, had the event not occurred. The same might be said of many medium sized or even large businesses.

The law considers damages under two main headings:

Pecuniary losses, which can include:

- loss of past earnings;
- loss of future earnings;
- loss of earning potential;
- expenses, including medical.

Non-pecuniary losses, which can include:

- past and future pain and suffering;
- past and future disability and loss of amenity;
- bereavement.

The law assesses damages under two further headings:

- **Special damages:** These are usually not difficult to ascertain and comprise loss of earnings and expenses incurred prior to trial, less any state benefits paid.
- **General damages:** These are much harder to assess and can include both pecuniary and non-pecuniary losses. The major difficulty lies in areas such as calculating possible loss of future earnings, taking into account matters such as promotions, pay rises, inflation and the like or compensating for the loss of limbs or bodily function.

Lord Denning has described this as 'an insoluble problem'. Generally, damages are awarded in these areas on the basis of a one-off capital sum calculated on the present value of future loss; that is, the capital sum, when invested, should yield an income stream equivalent to the present value of the employee's losses.

Despite these complexities, it is not difficult to imagine how, in serious cases, damages awards can amount to many hundreds of thousands of pounds – sums of money well beyond the capacity of many businesses and employers to pay from their own resources.

Employer's liability insurance

In order to ensure that employers do have the capacity to pay, the law insists that they take out insurance, and the legal framework is established by:

- Employer's Liability (Compulsory Insurance) Act 1969; and
- Employer's Liability (Compulsory Insurance) General Regulations 1991, and subsequent amendments.

The purpose of this legislation is to ensure that employers are able to meet any legal liability to pay damages, whatever they might be.

In health and safety matters, liability nearly always arises in the tort of negligence, and an action for damages can succeed if an accident or ill health can be shown to be the liability, but not necessarily the fault of, the employer.

Case law has established two important principles:

1 Employers must take employees as they find them. This is particularly relevant to health-related issues because an employer may find himself liable for the consequences of injury or ill health arising from a previous employment.
2 Employees may be negligent, but not wilfully so, and if they are, the employer may still find himself liable for the consequences of employees' negligent acts. However, the notion of 'contributory negligence' plays an important part in calculating the amount of any damages payable to a negligent employee.

In simple terms, the law takes the view that, when liability is established, it is the employer who is liable and he must compensate his employee by paying damages but the amount of such damages will be mitigated by the degree to which the employee's own negligence contributed to his circumstances.

Some important points of employer's liability insurance are given below:

- Employees must maintain a minimum cover of £2 m, any one employee or any one occurrence (it was increased to £5 million in 1998).

- All material facts known to the proposer must be disclosed at proposal stage.
- A valid certificate of insurance must be displayed at all workplaces.
- Cover extends to employees which usually includes labour-only subcontractors, particularly in the construction industry.
- Employees covered include residents of Great Britain and all those employed here for more than 14 days.
- Insurance must be provided under an 'approved policy', and in order to be considered as an approved policy, insurers are prevented from writing in certain onerous terms on the insured which, in turn, might prevent employees being paid.
- Policies remain valid even if the insured fails to comply with some of the terms of the policy.
- Insured are entitled to full indemnity. That is, full loss replacement, but no betterment.
- Loss mitigation is an implied term of most insurance contracts. This has become a particular issue since the introduction of the Management Regulations and in particular Regulation 3, Risk Assessment. If an employer's risk assessment has revealed that certain preventative or protective measures needed to be taken, and they were not taken, this may be enough to invalidate the policy. Hence loss mitigation in the form of improved house-keeping and risk-management procedures has been a major growth area since the implementation of the Management Regulations.
- Some employers are exempt, mainly national, regional and local government, statutory undertakers and NHS trusts. This is because they are held to be able to meet any legal liabilities from their own resources, which in effect are government funds.
- Cover extends to civil liability, that is, payment of damages and legal expenses, and also to legal expenses arising out of criminal proceedings. This does not extend to payment of fines or to jail sentences. It is not possible, nor is it legal to attempt, to insure against either.
- Employers may be guilty of a criminal offence if they fail to affect and maintain insurance for reach day on which it is needed, and if they fail to prominently display a certificate of insurance in each workplace. In each case, there is a fine of £1000 payable on conviction.

Other insurances

As the Derby City Council case in Chapter 1 showed, responsibility for health and safety management usually rests with more than one party, particularly in large organizations, complex business undertakings or extended contractual arrangements.

Often it is the very size and complexity of the undertaking or the arrangement which makes it difficult to clearly apportion responsibility and authority, and because of this, management structures can be difficult to define and can also be ineffective.

This is what was alleged in the *Herald of Free Enterprise* disaster, when it was said that management should have ensured that a safe work system was in place which would have prevented ships putting to sea with their bow doors open, and that such a procedure should have been included in the company's 'Ships Standing Orders'.

P&O's management was alleged to have been sloppy, and the company and a number of its directors were prosecuted, with many lawyers confidently predicting convictions and the establishment of the long-awaited precedent of 'corporate killing'.

However, the case fell apart before the prosecution had finished presenting it. P&O's own employees had testified that no-one had realized that there might be a risk that the ferry could put to sea with its bow doors open, and that was good enough for the judge against the background of existing case law, who found that the company and its directors and employees had complied with the law as it then stood, and ordered their aquittal.

Many people, particularly relatives and survivors, were dissatisfied with the verdict both in the case and in general. They have argued strongly that the test applied in that case should be replaced by new, much sterner tests, and *Law Commission report 237* (1998) now recommends two new offences against individuals, namely:

- reckless killing,
- killing by gross carelessness,

replacing manslaughter, and a new offence which can only be committed by a company, corporate killing.

In defending itself and its directors, P&O paid out £3.5 m in fees, and later sought to reclaim tax paid on them as services to individual employees and not the company. Again, the critics had a field day.

But what else was P&O to do? It had, after all, been found not guilty along with its directors and senior employees and it had a duty to them and to its shareholders to safeguard their interests. It seems also to have escaped the notice of the critics that the £3.5 m comprised some extremely high hourly and daily rates for lawyers, and such people are not noted for their ability to contribute to the design and implementation of safe work systems.

Against this background it is now becoming increasingly common for organizations to take out personal liability insurance for directors and senior managers with corporate health and safety responsibilities. It is not compulsory to do so.

Insurance: summary

In short, employers must contribute to the public insurance system enshrined in the national social security system.

They must also effect private insurance against legal liabilities within the framework established in the common law and under the Employer's Liability (Compulsory Insurance) Act 1969.

Employees may be entitled to benefits under the public system and damages under the private system.

Since the introduction of the Management Regulations in 1992, loss mitigation has become an implied term of most insurance contracts. Risk management and good housekeeping are the two major techniques of loss mitigation, and form an important part of compliance with the terms of employer's liability insurance policies.

3.7 Chapter summary and conclusion

Common law duties exist alongside statutory duties, but as more laws are made, so the importance of common law diminishes.

The HASWA/MHASWA framework is the main engine of health and safety management in the UK. It applies to nearly everybody in the country. Risk assessment plays a crucial role in the design of safe work systems, which must take into account the requirements of all other relevant regulations. If everybody fully implemented legal requirements, perhaps there would be no accidents and no injuries or ill-health. But they do not, and accidents occur which can have serious consequences.

Employers must accept the consequences and insure themselves to ensure they can meet any liabilities they may incur for compensating people injured or made ill at or by work.

4

Health and safety management: five steps

4.1 Introduction

Many people say that there is no such thing as 'health and safety', and what I take them to mean by this is that health and safety is not, or should not be, some separate activity taking place outside the main processes in the workplace.

The Health and Safety Executive says that there are five steps to successful health and safety management (IND(G)132 L 1992):

1 Set policy.
2 Organize staff.
3 Plan and set standards.
4 Measure performance.
5 Audit and review.

Setting aside health and safety issues, this is in fact excellent business advice on the strategy of running a business, any business. But it cannot be implemented to manage health and safety unless, and until, the main workplace process has been dealt with to a similar standard. Anyone attempting to deal with health and safety in the workplace or within a process as some separate issue soon finds that it is ephemeral, it has no substance, that it keeps drifting away unless it is firmly anchored to the reality of the work activity or process which is the main undertaking of the organization.

The separate approach usually leads to the generation of paper-based systems which add little or nothing to the efficiency of the process to which they are supposed to relate. They have much in common with paper-based quality assurance systems, and both are the first things to fall by the way-side when the pressure comes on. It can be instructive to consider quality alongside health and safety management because both have similar characteristics. There are three approaches to quality management within a business process:

- quality control (QC),
- quality assurance (QA), and
- quality management (TQM).

Quality control

Control is on the bottom rung of the quality ladder in any process. It implies that if there are no controls within a work process, quality is achieved by chance rather than intent. Controls must, therefore, be put in place to inspect and monitor the process, with a view to identifying poor-quality products once they are made, and finding and eliminating the causes. However, inspection and monitoring by themselves do not actually eliminate problems, they merely identify them when they happen and the end result is that the process manufactures defective products.

Quality assurance

Assurance is on the next rung of the ladder and is defined in ISO9000 (1994) as '**all those planned and systematic actions necessary to provide adequate confidence that an entity will fulfil requirements for quality**'. Quality assurance activities establish the extent to which quality is controlled in a process. The quality of the business, as distinct from its products, is measured along with its systems and procedures, thus giving assurance to the firm's management and its customers of the quality of its products and/or services. It is still necessary to implement procedures within the process to achieve quality products.

Total quality management

Total quality management is currently at the top of the quality ladder. Its proponents claim that it is relatively new, and that it is a philosophy. Oakland

(1989) produced the definitive work in the United Kingdom. The philosophy emerges on two fronts, first in defining quality, and second in achieving it.

There is no universally accepted definition of quality. In the business sense, the most widely recognized term is 'consistently meeting customer requirements', but some people say it is not achieved until the customer is 'delighted'.

There are three very broad statements used to described the philosophy underlying total quality management. They are:

- zero defects;
- get it right first time;
- continuous improvement all of the time.

Many issues flow from those three simple statements but taken at face value, TQM clearly cannot be achieved within a process that, for example, relies on quality control, for by definition quality control implies that there will be defects and things will not be right first time.

Similarly, a quality assured product or service will fail the third test, continuous improvement all of the time. Once a product or service is quality assured, it tends to stay the same, whereas to achieve continuous improvement, change must be constant. That is why, in practice, total quality management proves an impossible dream for most companies, for many are able to achieve customer satisfaction at quality assurance level or perhaps a little beyond and are therefore happy to stay there because their customers are satisfied. If this view is accepted, then it might be instructive to look at the minimum requirements for a quality assured product or service.

The starting point is therefore to set quality assurance within a framework. Figure 4.1 illustrates both a business process and a health and safety management structure. The latter is indistinguishable from the former.

Quality management in construction

The Construction Industry Research and Information Association (CIRIA) in its special publication 132 (1996) published the findings of its survey of experiences with BS5750, Quality Management in Construction. BS5750 was first published in 1979 and is now part of the ISO9000 series.

The survey makes distinctly lukewarm reading, particularly for enthusiasts of quality management systems, and clients of the construction industry. A total of 35 firms across the industry, including consultants and contractors, were interviewed in the survey. Most wanted to improve their firm's image and to generate better business opportunities.

Figure 4.1 Business process model/health and safety management model

Most were disappointed. Although all of them acknowledged that significant improvements were made to products and services, these occurred at a lower level than was expected. The survey identified two main reasons for this. First, expectations of extra work arising from quality assurance was set too high, and not as much work was won as was hoped for. Another way of looking at this is that clients did not value quality assurance as highly as expected. Second, most firms probably did not allocate enough resources to the vital internal auditing and management review procedures needed to consolidate quality management systems within the business. In other words, management commitment was not high enough.

Time and cost were also significant factors. On average, it took 14 months to design systems and a further 15 months to implement them, and costs varied from £1000 to £36 000. Performance improvements of between 40 and 87 per cent were claimed following certification, but excess paperwork was a constant complaint.

In summary, quality management systems in construction do seem to produce measurable benefits, but not as great as people hoped for or expect. It takes a long time to design and implement the systems and costs time and money to maintain them once they are installed, and unless they are maintained by monitoring and auditing, any benefits are quickly lost. Excess paper appears endemic.

It follows that the same lessons must apply to health and safety management systems. Continuous commitment from the top down is essential, and health and safety specialists must be developed and trained, not necessarily as some separate discipline, and have a say on the management team. Paper-based systems have an inversely effective relationship to efficiency.

4.2 Setting policy

Policy: 'a plan of action adopted by a person, group or government'.

Many organizations introduce their health and safety policy documents along the following lines.

This company fully accepts its statutory and common-law duties and responsibilities in relation to the safety and health of employees at work, and others not at work but affected by its work activities. We will do everything in our power to promote an active safety culture and are fully committed ... etc ... etc ...

Signed ————————————————————— Managing director

The problem with this approach is that the above is a statement, it is not a plan of action. It is usually followed by a series of further statements intended to address various aspects of HASWA/MHASWA which collectively comprise the policy. Such documents usually end up gathering dust in the back of a filing cabinet.

Most organizations have two basic objectives:

■ survival, and
■ making money.

It is very difficult to achieve both. Over 90 per cent of new businesses fail in the first year (source: DOE). Fewer than 5 per cent of businesses last five years and about 1 per cent survive for more than 10 years. There are many reasons for this low business survival rate, but the main reasons are to be found in the areas of human aspirations, behaviour and expectations.

Basically, most of us think we are more talented than we really are, we usually over-rate the standards of our own behaviour and these factors lead into unrealistic expectations of what we are entitled to in life. Many of us are also basically lazy, in some way or other. It is widely recognized that most managers do not like planning; many do not like working to plans, either their own or those prepared by others. Some would say this is not laziness, it is some other human condition. Fair enough. But whatever its cause, the effect – the lack of planning – is the same.

The purpose of planning is to exert control, so without a plan there can be no control and when there is no control, businesses fail. The first effect is lack of control and the ultimate effect is failure of the business. It is exactly the same with health and safety. Lack of planning leads to loss of control which results in accidents, injuries, ill health and for businesses, the fatal disease of loss through waste. This is perhaps the most persuasive business argument for not dealing with health and safety as some separate issue. On its own, it is perceived to be burdensome and of no value to the business. 'Oh its only health and safety, I can't be bothered with all that . . .' and so on. We have all heard the call of the dinosaur.

But the point is that health and safety is not some separate issue because it is part of any business process and must contribute to the wider objectives of the business. The problem is, that if the business has only two objectives – survival and making money – that will almost certainly not be wide enough to sustain the business in trading into its second year and beyond.

In many ways it is a pity that the legislators allow the existence of separate health and safety policy documents, for that can encourage the view that it is some separate issue and can be dealt with as such. There is a school of thought that says the statutory duty should be to plan your business properly, ensuring that the business plan deals with health and safety in a way that contributes to the broad business objectives. At this point, it might be sensible to consider what broad business objectives can be effected by health and safety planning:

- corporate strategy;
- corporate planning;
- product research and design;
- manufacturing/processing/operations;
- human resources;
- IT and information management;
- corporate finance;
- marketing and selling;
- product liability and insurance.

The legislators might reply that if these matters were to be brought within the remit of health and safety law, and it became a statutory duty to *'plan your business, so far as is reasonably practicable'* then our courts and prisons would be overflowing with miscreants within a matter of days. I am sure they would be right. So, whilst it is probably true to say that the policy introduction at the start of this chapter meets the statutory requirements of HASWA, it falls a long way short of the minimum to be expected of a competent plan of action.

There are a number of well-known and well-developed business and project management techniques which are discussed more fully under the heading of 'planning', but in outline MOGSA methodology is a widely recognized technique in common use across business and industry. It is usually used at the strategy stage of projects or new enterprises for constructing a model of the proposed project or enterprise and it works in the following way:

Mission
Objectives
Goals
Strategies
Actions

Mission   our mission is to eliminate waste arising from accidents and ill health from our business process.

Objectives achieve zero notifiable accidents year on year.

Goals   reduce the cost of all unplanned events resulting in loss, damage or the loss of a business opportunity to less than 5 per cent of turnover this year and less than 2 per cent of turnover next year.

Strategies quality assure office processes.
ensure office/site interface links.
quality assure site processes.
improve productivity by 10 per cent every year.
eliminate manual handling of materials.
implement benchmarking.

Actions   implement JIT materials delivery.
eliminate material storage on site.
make CPD compulsory for office stage.
make CITB training compulsory for operatives.
improve IT facilities.
improve planning and programming techniques.

The above is an example of the application of MOGSA methodology to the writing of a policy statement for a construction company committed to implementing total quality management through the construction processes under its control. Again, the health and safety issues are indistinguishable from good business practice, and, although the list is far from exhaustive, it covers many of the best-practice issues relevant to many businesses, not just building companies.

Policy: summary

Policy at business process level is established in best-practice organizations by the use of techniques such as mission statements. It is usually not good enough to make some bland statement. The lesson applies to health and safety management as well as to the broad business process. Very senior management must make a firm commitment designed to pull an organization towards a specific goal, rather than push it towards some unspecified target.

4.3 Organizing staff

Organize: 'to plan and arrange systematically'.

The first hurdle to be overcome in organizing staff is to understand what is meant by 'staff', on an organization-specific basis, and in construction, on a project-specific basis. The term is generally applied to persons directly employed by the organization for which they work. Even then, there are frequently hierarchical divisions between managers, staff and workers within organizations, and then a further hierarchical split involving temporary workers, labour-only subcontractors and the self-employed. Again, this is particularly evident in construction.

In the UK, the workforce in 1995 stood at about 15.72 m and across all industries temporary workers comprised 5 per cent and self-employed persons comprised 13 per cent of the workforce (source: UK Labour Force Survey, DOE). In construction, the figures stand at 6 per cent and 45 per cent respectively.

In other words, construction uses about two-and-a-half times as many self-employed persons as most other industries. About half of the people engaged in construction are not, or do not consider themselves to be, directly employed by the organizations for which they work, and the same attitude prevails with employers. The tendency is for each group, employers and the self-employed, to hold each other at arm's length for various reasons.

Employers use the self-employed mainly to help them to gain competitive advantage in the marketplace and to avoid the business risks of employing people directly. It is commonplace for employers to believe that they can also avoid the health and safety management risks attaching to self-employed persons.

Self-employed persons usually wish to retain flexibility in who they work for, variety in what they do, and above all to benefit from an anticipated higher income and the tax advantages of self-employment.

Frequently, this kind of relationship is not arm's length at all. It is artificial in the sense that the self-employed continue working for the same employer for as long as work lasts, which can be for years, sometimes on fixed-price or lump-sum contracts and sometimes on hourly time charges, and nearly always on a labour-only basis.

In recent years, the Inland Revenue has taken a fresh look at this situation, particularly in construction, and so has HSE across industry generally. In 1997 the Inland Revenue launched a major clamp-down on self-employment which they say (March 1998) has resulted in about 147 000 workers returning to direct employment. Case law arising from the tax clamp-down has established that exact employment status is determined by the true circumstances of each case, taking account of matters such as whether or not the worker provides his own tools and materials, what business risks they take, and crucially the amount of control each worker has over what he actually does. The courts rule on the basis of the facts, which is not always the same as the positions set out in contracts.

To the extent that self-employed persons do not provide tools or materials, take no business risks (for example, are paid by the hour or are paid for correcting their own mistakes or defective work), work for the same organization for long periods, and do what they are told when they are told, then the courts have rejected self-employed status, holding on the basis of the facts that such people are employees. This has enabled the Inland Revenue to recover from employers substantial amounts of tax unpaid by self-employed people who have worked for them.

Similar precedents have been set in health and safety matters, under the general principle that the duties under the law fall on the organization or person in the best position to discharge them. However, the crucial issue under health and safety law is not so much to do with the provisions of tools and materials, the way in which people are paid or the acceptance of business risk, although these factors can be relevant; instead, it relates to who is actually *controlling* the workplace activity.

So the starting point in organizing to manage health and safety in all workplaces, particularly construction, is to truthfully answer two questions:

1 What is meant by staff?
2 Who is actually controlling the workplace activity?

In most cases, staff will include all workers, howsoever employed, participating in the workplace activity, and the main employer will be controlling the workplace activity. Unless there are exceptional circumstances, it is probably best practice for employers to accept these answers as fact and to make appropriate organizational arrangements.

Control

If control of the workplace activity is the key issue, then the key element of control is planning. Without planning, there can be no control and there is no point in planning unless you intend to control. A business cannot be sustained and money cannot be made if planning is absent.

The two prime reasons for exerting control are:

■ to enable managers to establish that achievable objectives have been set; and
■ to assist them in taking decisions relating to those objectives.

In order to implement effective planning and control, three other elements must be present:

■ communication,
■ co-operation,
■ competence.

Communication

There is no point planning if people are not told what is planned. Communication is, therefore, an essential part of control and takes place in three recognized ways within organizations: verbal, written and observed.

Verbal communication includes:

■ meetings, one-to-one discussions
■ team briefings
■ quality circles
■ tool box talks.

Written communication includes:

■ policy statements
■ organization charts

- performance standards
- risk assessments
- posters
- newsletters.

Observed communication includes:

- Regular inspection tours
- senior management involvement in meetings
- setting an example
- joint consultation meetings
- presentations/training sessions and workshops.

Once more, all of these issues are simply part of good business practice in running any undertaking, and the health and safety bits are more or less indistinguishable.

In all organisations, there will also be an element of external communication needed to deal with issucs such as:

- reporting accidents and ill health;
- dealing with statutory paperwork;
- interfacing with HSE;
- interfacing with information services;
- liaison with statutory undertakings and bodies.

Co-operation

There is no easy way to ensure co-operation between individuals and organizations. Coercion is rarely successful. Encouragement and involvement are nearly always more successful, but it is a matter of degree over a timescale. No matter how skilfully people are handled, their commitment levels vary.

It is instructive to note that both HASWA and the CDM Regulations place duties on people to co-operate and make it a criminal offence if they fail to comply, in the discharge of health and safety responsibilities.

Most managers recognise that the key to co-operation is 'ownership', but difficulty is nearly always experienced to a greater or lesser extent in getting people to accept ownership of issues. This is particularly so where there are different categories of worker, different types of employee or employment, and may also go some way towards explaining why Japan has been successful over the past 50 years in dominating manufacturing industries.

In most Japanese workplaces, there is no hierarchical structure. Management and workforce wear the same clothes, eat in the same canteen, park in the same carpark, undergo the same training and, from time to time, share the same tasks. Japanese philosophy towards work and workers is surprisingly simple, and is to the effect that people generally do not mean to make mistakes but do so inadvertently and regularly. Inadvertent mistakes occur through forgetfulness, so a common Japanese exhortation is:

A person is an animal that forgets. Learn more than you forget.

As always, the Japanese have a philosophy which embraces co-operation in the business process. It is called poka yoke (yoken = to avoid; poka = inadvertent errors, and it is based on respect for the intelligence of workers, and the conviction that it is unacceptable to produce even a small number of defective products. It is a complete reversal of traditional western thinking, which accepts the fact that defects can and do occur; instead, the Japanese view is that defects do not occur, they are *made* inadvertently.

Dr Shigeo Shingo wrote the definitive book *Poka Yoke: Improving Product Quality by Preventing Defects*, which sets out in detail the tools and techniques of co-operation used in achieving poka yoke, but the philosophy requires that the design development process must incorporate the experience and skills of the production workers for they are in the best position to discover design elements that cause difficulty and add little, or no value, to the process or product.

Any efficient business process must, therefore, find effective ways of integrating the activities of designers and workers, designing with manufacturing. This is fundamental to all business and workplace processes and is generally well recognized in most major industries. Again, however, construction is different in so far as it is the only major industry that separates product design from product construction in quite the manner that it does. The detail of this is discussed more fully in later chapters, because it can have a major impact on the way in which health and safety is managed on building projects.

Figure 4.2 illustrates a typical product manufacturing process cycle. Figure 4.3 illustrates a typical construction process cycle. The point of interest is the separation of design from construction. Note also the sequence involving defects correction.

Given that a co-ordinated design/manufacturer works process model is achieved, then the following tools are available to assist in communication:

- training schemes;
- consultation procedures;
- problem-solving teams;

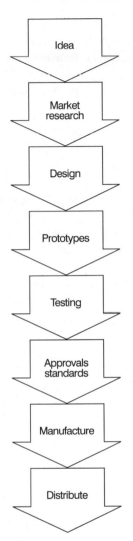

Figure 4.2 Typical product manufacturing process model

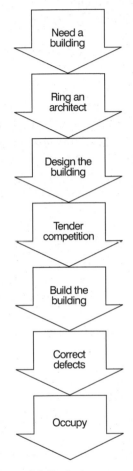

Figure 4.3 Typical construction process model

- hazard reporting schemes;
- suggesting schemes;
- quality/safety circles;
- safety representatives.

Of the above, HSE actively promote the use of **safety representatives**, for the valuable contribution they can make to improved communication between management and the workforce using the catalyst of health and safety. There is no formal role or qualification of safety representatives, but it first came to prominence with the implementation of the **Safety Representatives and Safety Committees Regulations 1977**, and its equivalent, the **Off-shore Installations (Safety Representatives and Safety Committee) Regulations 1989**.

Part of the value of safety representatives is that they are part of the workforce and remain as such, but undertake additional training in health and safety management issues, and are ensured adequate time to put their training to use in the workplace.

The writer regularly sees the work of safety advisers and committees in industry, particularly in manufacturing and process engineering. They are rarely witnessed on construction sites, although they can be seen from time to time on large or prominent projects. The writer has never encountered a safety representative as such in the organizations of consultants engaged in construction.

Experience has shown that safety representatives and committees can become embroiled in conflict. This is usually rooted in the fact that in many hierarchical organizations, the representative or committee can be the first or only direct point of contact between management and workforce. It provides employees with opportunities to 'have a go' at management which have previously been denied to them. It needs managing, usually in the form of an independent facilitator or external adviser, but must simply be regarded as part of the experience of implementing a health and safety culture, not an excuse for abandoning it.

Competence

In simple terms, the law requires all people with health and safety duties to be competent to discharge them. Regulations 4 and 6 of MHASWA are important to all employers and Regulations 8 and 9 of CDM are important in construction.

Because it is, therefore, something of a moving target, the first thing any organization needs to do is to define competence for the following categories:

- the organization itself;
- the employer;
- employees;
- the self-employed.

The purpose of this exercise is to develop a training programme to achieve compliant levels of health and safety competence.

The first step is to assess existing competence levels. This can be done by means of questionnaires and/or interviews with the affected parties, but its purpose should be made clear. In one large metropolitan borough council in the north west, the writer encountered a major staff/management dispute which had originated when the latter instituted an assessment programme which was not explained to staff in advance. They mistook it as some preliminary procedure intended to lead to job losses and redundancies, when in fact, the opposite was the case.

Having benchmarked competence levels, the next step is to design and implement a training programme for each of the above four categories. Figure 4.4 illustrates the process.

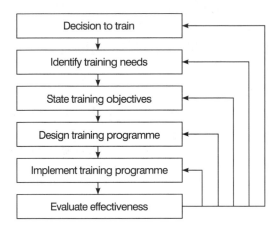

Figure 4.4 Model of training process

Training needs generally fall into three broad categories:

- **Very senior management (policy makers)**: Training programmes for this level need to be directed at broad strategic and operational needs of organizations. Very senior managers need to know enough to understand and comply with their own individual duties and to appreciate what their organization must do to achieve compliance, and thus to be in a position to authorize training for other levels.

- **Management and supervision (implementers)**: Training programmes at this level need to deal with leadership and communication skills, planning, benchmarking and monitoring systems, risk assessment and safe work system design and training and problem resolution methods.

- **Workers (doers)**: Training programmes at this level need to provide a broad overview of the health and safety system in the UK, and must then address details of the relevant legislation, HSE guidance, and the company's safety management systems.

Training is one important aspect of competence, but there are a number of others which include the following:

- the organization's resourcing policy;
- external recruitment policy;
- internal placement policy;
- calibre of people employed;
- manner in which people are employed;
- fitness and health of people;
- imported and acquired experience.

Resourcing policy

This embraces the debate between directly employed or self-employed people. In the writer's view, there is no distinction to be made between the two as far as health and safety law is concerned, and organizations are best advised in most circumstances to treat both categories the same. This does not preclude employers requiring the self-employed to contribute towards the cost of any specialized training provided for them.

External recruitment

The debate here usually centres on experience or youth. Experienced people cost more, and can bring with them a certain amount of baggage in the form of preconceived ideas or set patterns of work and behaviour. Youth is usually less productive, initially at least, and can bring attitude, aptitude and commitment problems. Young people usually cost less in wages, but this must not be confused with being cheaper to employ.

Internal placement

Many companies promote from within, or fill job vacancies internally first. This can be a barrier to raising health and safety competence, particularly if safety representatives have a low existing profile. Very often external recruitment can significantly improve an organization's performance through the introduction of new ideas and fresh ways of thinking.

Calibre of people employed

In an age of equality and fairness, this is perhaps the great unheard debate. The old adage that 'if you pay peanuts you get monkeys' is as true today as it ever was and probably always will be. By calibre, is meant the attitude, aptitude and ability of people towards their work. These are not easy qualities to assess, but they dictate the value and work of any person within an organization. One manager the writer knew used to judge building workers on two main criteria, whether or not they smoked and the way they walked. If they didn't smoke and had a good lively walk, he would employ them on the grounds that they probably had some life in them, and could achieve acceptable productivity and quality standards.

Manner in which people are employed

It can make a big difference to people whether or not they are directly employed, on a temporary or fixed-term contract or self-employed. Not only will it dictate how they are paid, but also the way they feel about the job they do. This will impact on commitment and motivation levels and on their approach to 'ownership' of health and safety issues. The debate is not one-sided. Self-employment for most people is a choice and has not been thrust upon them.

Fitness and health of people

Most active sports men and women must regularly endure the jibes and taunts of colleagues who, for whatever reason, seem to feel that physical exercise is to be avoided at all costs.

Fitness and health is of vital importance both in life and in the workplace. A reasonable level of fitness can be achieved and maintained by everybody using about three 20 minute exercise sessions a week; more than that is a bonus. It impacts on everything we do, but particularly on mental and physical health and well being. Generally, fit people are more alert and aware, they react more quickly, they are stronger, have more stamina and look better and healthier. They cope with stress better and have more resistance to occupational illness and disease resulting in less lost time due to illness and injury.

Increasingly, organizations are turning to active fitness programmes for workforces, particularly managers, to improve performance. Yoga courses are particularly suited to both fitness and stress relief within workplaces.

Again, the construction industry is a long way behind other industries in these areas.

Imported and acquired experience

Few workplace activities remain unchanged on a long-term basis. Indeed, in a TQM environment, it is not possible for this to happen; change must occur. Consequently, experience levels must change along with the process, and new skills and experience acquired. These may embrace the use of new plant, equipment, materials, techniques or processes where imported skills are needed before they can be acquired in-house.

Employers have duties to protect the health, safety and welfare of employees and must be competent to do so.

Employees have duties to themselves and to those who may be affected by their work. Self-employed workers have duties to protect the health and safety (but not welfare) of people not in their employment who may be affected by whatever work task they are undertaking – this includes themselves.

Organizations have the same responsibilities as employers, and modern thinking is leading many organizations to actively encourage employees to take a much more active interest in their own health and welfare. For these reasons competence is a moving target which is discussed further in later chapters.

4.4 Planning and setting standards

Concept

Figure 4.5 illustrates the concepts underlying most work processes. The ultimate aim of most workplace activity is to provide customers with products or services which meet their needs. Therefore, customer needs are the minimum standards which all business processes must meet. Total quality management philosophy says that the maximum standard is customer delight.

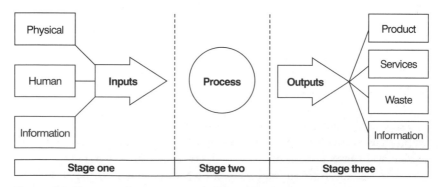

Figure 4.5 illustrates the concepts underlying most work processes.

Accordingly, management must decide in concept terms, at which ends of the spectrum it will set performance standards, which must include health and safety management standards.

Performance standards are used to control the flow of resources and information through the process. Consequently, there must be clear performance standards in place at each of the three stages of the process, to achieve both overall control of the process and of the health and safety management standards needed alongside and within it.

Inputs: stage one controls

These apply to physical and human resources and information taken into the process. The health and safety management objective is to identify and minimize hazards entering the process through inputs.

Performance standards should be applied to the following:

1 Physical processes to be used in the process including:
   ■ materials and substances
   ■ plant and equipment
   ■ premises, buildings and structures
   ■ utilities
   ■ other businesses
2 Human resources to be used in the process, including:
   ■ direct employees
   ■ self-employed workers
   ■ consultants
   ■ contractors.
3 Information to be used in the process, including:
   ■ compliance with legal requirements
   ■ compliance with contractual requirements
   ■ relevant British Standards and codes of practice
   ■ performance standards to be met
   ■ functional standards to be met
   ■ technical standards to be met.

Process: stage-controls

These apply to the process itself and to its components. The health and safety objectives are to create a health and safety management culture alongside the work process culture at at least the same level (e.g. QA, QC, TQM or other) and to minimize risks occurring in the process activity.

The process usually comprises the following components:

1 Premises to be used in the process itself, including:
   ■ access, egress and parking
   ■ the workplace itself and its physical characteristics
   ■ the workplace environment
   ■ welfare facilities
   ■ fixed plant and equipment, e.g. electrical installation
   ■ fabric condition and maintenance requirements.
2 People engaged in the process itself, including:
   ■ managers
   ■ supervisors
   ■ directly employed people
   ■ self-employed and temporary workers
   ■ external consultants and contractors
   ■ health and safety specialists.
3 Procedures used in the process including:
   ■ designing of products
   ■ selection of work processes
   ■ specification of jobs
   ■ design of safe work systems
   ■ all workplace activity.
4 Plant and materials used in the process including:
   ■ wholly owned or hired-in plant
   ■ handling transport and storage arrangements
   ■ operating and maintenance procedures.

Outputs: stage-three controls

These apply to the organization's products, services, waste or by-products and the information it produces to sell or explain its products and services. The health and safety objective is not to export risk through outputs to other organizations.

Products can include the following:

■ research and development information
■ packaging, labelling and storage information
■ usage instructions
■ health surveillance needs arising from use
■ delivery, transport and off-loading instructions
■ installation, setting-up and testing needs
■ cleaning, maintenance and refurbishment needs.

Services can be either stand alone or provided as part of a package and can include:

- the way in which the services is to be provided
- timescale for provision
- rates and dates of delivery
- personnel involved in supplying.

Waste arising from the process can include:

- outputs to environment or atmosphere
- disposal issues.

What standards to set

The concepts and outlines described above apply readily to homogenous processes taking place at a single large site or to a single coherent activity. For example, most car manufacturers, or organizations making components for the automotive or other industries, would have little difficulty in adapting the broad principles outlined above to suit their workplace activities.

Basically, the concepts have been evolved against the assumption of 'steady state' activity and an unchanging location. Clearly, this applies to most of business and industry, but equally clearly, it gives rise to particular difficulties when considering construction. Before turning to construction, there is a final point to be made, which is, what performance standards are to be set?

Is there a general set of standards that can be applied across all processes at each of the three stages? The truthful answer is, probably not, but there is a time-honoured trio of performance assessment criteria which most people use, consciously or unconsciously, in making business decisions. They are:

- **Time:** programme
- **Cost:** price
- **Quality:** performance.

Apart from the fact that they are the most important things on people's minds in most undertakings, all three can be quantified to a reasonable degree of accuracy to enable achievable targets to be set, and can be used in most business applications. Many people argue that safety should automatically be added to the list, others say it should not be included as a separate issue.

An interesting feature of performance standards is the way in which they impact on each other. If time cost and quality are chosen as the standard performance setting technique for a business process, then Figure 4.6(a) illustrates what happens when equal priority is given to each of the three criteria within the business process.

However, this very rarely happens. One of the criteria is nearly always more important than others and is, therefore, given more weighting. Figure 4.6(b) illustrates the impact of this.

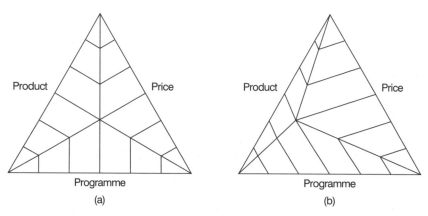

(a)

(b)

Figure 4.6

Example

Time, cost and quality standards could be used as follows. A client, perhaps a manufacturer of parts for the automotive industry, might wish to start a new manufacturing process on a greenfield site for a new product.

At Stage 1, time, cost and quality could form the framework for:

- acquiring suitable premises, either new-build or existing
- recruitment of the workforce
- training of the workforce
- designing the product or service.

At Stage 2, time, cost and quality could form the framework for:

- operating and maintaining the premises
- acquiring process plant and equipment
- acquiring materials for use in the process
- designing the work sequences
- maintaining workforce competence levels.

At Stage 3, time, cost and quality standards will be crucial in:

- the selling price of the product or service
- after sales issues.

Planning and setting standards in construction

In an homogenous process it has been identified that in concept terms there are three stages to be considered, namely inputs, process and outputs.

Various controls are needed at each stage, and assessment criteria are needed against which to measure performance. The classic trio are time, cost and quality. These three are particularly useful because they are widely understood and can be relatively easily quantified.

Construction is a process that comprises two main and separate activities:

- designing and managing
- building and operations on site.

There are six separate procurement routes in common use, and available to clients of the construction industry. The routes are:

- traditional
- design and build
- design and manage
- management contracting
- construction management
- engineer, procure, construct (turnkey).

The detail of these routes and their impact on health and safety management arrangements on construction sites are examined in Chapter 5. An important point, however, is that the CDM Regulations seek to ensure that health and safety is managed continuously *across the lifecycle* of construction projects. An illustration of project lifecycle is described in Figure 4.7.

In practical terms, the design and build separation in construction can be illustrated by the model in Figure 4.8. Essentially, the model is showing two separate processes, designing and building, each with its own three-stage cycle.

Construction is said to be an 'iterative' process, that is, before a structure can be built, it must be designed. The former cannot start without the latter, and any defects or non-conformances in the designed product flow straight

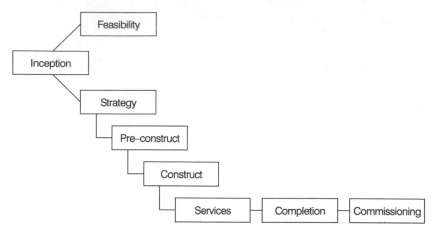

Figure 4.7 Stages of construction project (CIOB)

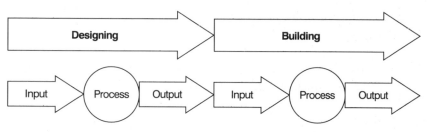

Figure 4.8

into the built product. It may seem strange to those outside the construction industry, but this seemingly self-evident point is not widely understood within construction, because there is little understanding as yet of construction as a process.

> Process management is not well understood within the construction industry. Indeed, it is a subject which will have little immediate resonance in an industry which for almost 20 years has been organised in disbursed, traditional structures generally constrained by finely detailed contracts and regulations.
>
> (Department of the Environment, Transport and the Regions, 1997)

Towill (1997) applied a business systems engineering approach to a study investigating ways of improving construction productivity. Table 4.1 illustrates his findings.

Table 4.1

Cause of wasted time on site	Effect (percentage total waste)
Re-work due to drawing changes	40
Re-work due to job done wrong	30
Inefficient work methods	15
Wasting time due to defective materials	5
Changed customer requirements	5
Other sources	5

In contemplating this data, the obvious question to be asked is 40 per cent of what? The data was based on the difference between planned performance and actual performance on site. In other words, 40 per cent of time wasted on the sites comprising the study was caused by changes in design information impacting on those carrying out the building work.

Planning and setting standards: conclusion

Planning and setting performance standards in construction can follow best practice in other industries, but it must be recognized that there are two distinct and separate aspects to construction, namely designing and managing, and carrying out work on site.

Although the concept of health and safety management for designers and managers is the same as for builders, the details vary considerably and to fully understand the differences, it is necessary to examine the six major procurement routes in detail. This is done in Chapter 5.

Common ground

The central requirement of any successful business process is a rigorous risk management methodology. In the business sense, risk is described in Figure 4.9.

In commerce, risk is generally thought of in the following way:

> Exposure to the possibility of economic and financial loss or gain, physical damage or injury, or delay as a consequence of the uncertainty associated with pursuing a particular course of action
>
> (Chapman, 1991)

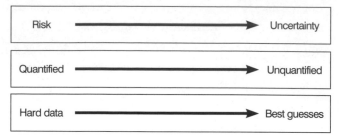

Figure 4.9

Modern risk management techniques in business and commerce have their origins in the US insurance industry in the 1940s. Risk could be assessed and insured using the formula:

Risk = Probability of event × Magnitude of loss/gain

Uncertainty usually describes situations where it is not possible to attach a probability to the likelihood of an event occurring, and is therefore uninsurable.

When comparisons with health and safety techniques are made, there are, perhaps not surprisingly, close parallels. **Hazard** is the potential for an unplanned event to occur which might cause harm. **Risk** is the likelihood and severity of the unplanned event actually occurring, and causing harm.

Performance standards are needed to control risks arising from hazards occurring in workplace activity. Therefore, a robust risk assessment procedure is an essential feature of performance standards needed to control construction. Risk assessment must include the following features:

- hazard identification;
- risk assessment;
- risk control;
- maintenance of control measures.

Separate and distinct performance standards, and therefore risk assessment procedures, must be applied to:

- design and management of construction;
- carrying out work on site.

4.5 Measuring performance

Business performance

In the broad business environment, 13 separate ways have been identified in which projects can perform. Figure 4.10 illustrates the 13 ways. In order to understand how a project is performing, it is obviously necessary to monitor and measure performance, and current project management thinking postulates that both **time** and **cost** are monitored by the management structure applied to the project.

MOGSA methodology models the project (see p. 57).

It is then necessary to choose an appropriate organization structure to implement the model. There are three types available:

Functional structures

This is the classic management pyramid based on specialization, line and staff relationships, commensurate authority and responsibility, defined areas of control and so on. Its basic strength is the concentration of like-minded

Example	BCWS	ACWP	BCWP	Condition	Key
1	1000	1000	1000	☺	Planned = actual
2	1000	800	800	☺	On budget, behind time
3	1000	800	1000	☺	On time, ahead of budget
4	1000	800	1200	☺☺	Ahead of time and budget
5	1000	1000	800	☹	Behind time and budget
6	1000	1000	1200	☹	Ahead of time, behind budget
7	1000	1200	800	☹☹	Badly behind time and budget
8	1000	1200	1000	☹	On time, behind budget
9	1000	1200	1200	☺	Behind time, over budget
10	1000	800	600	☹☺	Behind time, on budget
11	1000	600	800	☹	Ahead of time, over budget
12	1000	1200	1400	☹	Ahead of time, behind budget
13	1000	1400	1200	☹	Behind time, over budget

BCWS = Budgeted cost for work scheduled
ACWP = Actual cost for work performed
BCWP = Budgeted cost for work performed

Figure 4.10

professionals within each discipline which can often result in high levels of technical excellence. However, problems can be encountered as a result of elitism and concentration on inward looking goals rather than overall project objectives.

Pure project structures

This is a single purpose structure established as a self-contained team under the direction of a single manager as a one off exercise and the team normally contains all of the specialist skills necessary to undertake the project. Singleness of purpose and unity of command are two of its strengths. One significant weakness is that facilities and resources may be duplicated.

Matrix structures

These are hybrid structures which attempt to optimize the strengths and minimize the weaknesses of functional and pure project structures. It is usually conceptualized as a vertical functional structure with a super-imposed horizontal project structure.

Its great strength is that functional specialisms allow for greater co-ordination to be achieved, but its major weakness is usually duality of command resulting in reporting to functional heads in one direction and to the project manager in the other.

Each type of organization structure requires the following features for full and successful implementation:

- Work breakdown structure;
- Organization breakdown structure;
- Cost breakdown structure;
- Work package activities;
- A coding system based on the above;
- a WBS (work breakdown structure) dictionary to hold the data;
- computer literate people;
- an earned value mechanism.

The purpose of a project management structure is three-fold:

- to enable managers to plan the project and exert control;
- to enable managers to monitor planned progress against actual progress (the 13 conditions) against a range of performance criteria (usually time and cost);
- to enable corrective action to be taken where need is identified.

Earned value is a concept worth explaining further, and is illustrated simplistically as follows. On a multi-storey brickclad building, experience shows that a brick layer might lay 100 bricks an hour at ground floor level, 90 an hour at first floor level, 80 at second floor level and so on. In other words, performance drops off with height, not because the bricklayer is less productive, but because more non-productive elements are introduced into his work. He has further to go to his workplace, it takes longer to get materials to him, weather conditions at height may not be the same as at ground level (wind can be a problem).

So, any project management system, to be effective, needs an earned value mechanism to enable variances to be monitored from planned performance. When variances are noted, the following strategies are available in attempting to regain control over construction projects.

- re-negotiate dates
- plan to recover time later
- narrow the scope of the project;
- deploy more resources;
- substitute materials;
- alternative sourcing;
- partial handovers;
- offer incentives;
- demand compliance;
- mutual assistance and co-operation.

The above is a thumb-nail sketch of current thinking in project management, with particular reference to construction projects. Figure 4.11 illustrates the

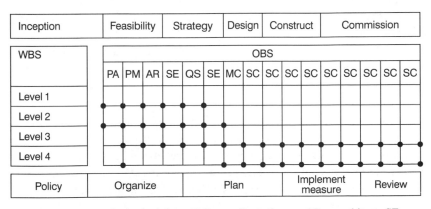

Figure 4.11 PA = principal adviser, PM = project manager, AR = architect, SE = structural engineer, QS = quantity surveyor, SE = services engineer, MC = main contractor, SC = sub contractor

concept of the WBS/OBS matrix and for further reading on the topic, the CIOB Code of Practice for Project Management is widely accepted as the standard reference work. The observant reader will note that the project management arrangements apply to the whole project, including designing, building and using at least up to the commissioning stage. The point has already been made that designing and building are *de facto* separate activities in construction.

Recent developments

Until 1994 it is probably true to say that there were two separate and distinct standards applied to health and safety management in construction.

Considerable emphasis was placed by all interested parties on work activities on construction sites, whilst a good deal less emphasis was placed on work activities taking place in the offices of consultants engaged in designing projects. Scandalously, hardly any activity at all was going on in universities and colleges offering higher and further education courses in the built environment.

One effect of this was that the performance of the contracting side of the industry is quite well known. After agriculture, it has the worst safety record of all industries; notifiable accidents cost about 3 per cent of turnover every year, APAU (Accident Prevention Advisory Unit, HSE) studies show that a loss control approach reveals the true cost of accidents to be at least 8 per cent of contractors' tender prices, and a worker spending his work life on construction projects has a 1 in 75 chance of being seriously hurt.

Another effect is that no-one has any real knowledge of the performance of the construction design and management industry, and therefore, the influence that designing and management arrangements have on the safety of those engaged in constructing, cleaning, altering, maintaining or demolishing buildings or structures.

The CDM Regulations, introduced in 1994, are expected to change this in time, but at present there is still little, if any, real evidence of the impact and cost of unsafe design and management arrangements in construction.

A third effect is that there is no knowledge of the impact made by educators or trainers in contributing to improvements in health and safety in construction. Sadly, it has to be said that the contribution of the education and training industry is probably a negative one, because, for the most part, it stubbornly refuses to address its responsibilities fully and properly within its built environment courses. This results in graduates being sent out into business and industry who have no proper training in the health and safety system in the UK, and therefore, little idea of their roles and responsibilities.

Lessons to be learned from the business environment

All workplace activities require a proper planning and monitoring system, which allows managers to understand exactly how the activity is performing in terms of planned versus actual, with the objective of taking appropriate corrective action in timely fashion. Therefore, any such system needs to be forward looking and capable of predicting future trends or events on the basis of recent past performance.

Application to health and safety management generally

Clearly, systems to measure and record health and safety performance should be part of, and sit alongside, existing workplace reporting systems if possible. Accordingly, there will be a number of different levels of workplace activity at which measurement can take place, and there will be different types of measuring and monitoring systems.

Types of monitoring system

'Successful Health and Safety Management' (HS(G)65) identifies two types of system: active and reactive.

Active monitoring systems

Essentially, these are intended to be forward looking and to measure success, not punish failure. Their equivalent in the business line is the WB/OB structures used for monitoring planned performance against actual. The major issues for managers rest in deciding what to measure.

There is always a cost/benefit trade-off to be made in all reporting systems, and the important thing is to choose a measuring level and system where the benefits flowing from the data outweigh the costs of collecting it.

The type of quality control/assurance/management system in use will materially affect what is measured and how it is measured.

The following are the areas of the business process which can benefit from active monitoring, and which can be monitored:

Process inputs:
- Physical resources
- Human resources
- Information taken in.

Process:

- Premises used
- People employed
- Procedures implemented
- Plant, materials, equipment utilized.

Process inputs:

- Products exported
- Services exported
- Waste produced and exported
- Information disseminated.

How to monitor?

This depends on management attitude to quality in the business process. In a quality control or quality assured environment, monitoring will generally be by inspection of some sort. In a total quality management environment, it will be part of a poka yoke or similar approach, which again involves inspection. The important point is that the inspection occurs before risks eventuate, and is part of a preventative process.

Inspection can be split into three categories:

- **Statutory**: That is, the need for inspection arises from a set of regulations, for example: pressure vessels; lifts; cranes; chains, ropes and slings; excavations; scaffolds; and exhaust ventilation and so on.
- **Contractual**: for example, arising under an insurance policy
- **Process**: arising from procedures laid down to ensure efficiency in the business process.

As a minimum, inspection should be carried out by competent people, and should be based on pre-determined schedules. Full records should be kept of all inspections and the outcome should be communicated to interested parties, which will always include management, safety representatives and the workforce.

If inspection reveal non-conformances in the process, there must be follow-up action. This is merely good business practice, but there is no point in producing inspection records if they are not acted on and simply left to gather dust.

Just as the WBS/OBS/EV (EV = Earned Value) mechanism alerts project managers as to which of the 13 progress states have been achieved, so active monitoring systems should be designed to fit into whatever process management system is in use.

Reactive monitoring systems

Again, there are three components, compulsory, contractual and process driven, but the important difference is that reactive monitoring, as the name implies, occurs after the event, for example in response to a risk eventuating, and includes:

- reporting of accidents and dangerous occurrences;
- ill health;
- loss or damage to property, plant and equipment;
- non-conformances in performance or quality standards;
- hazards.

Reactive monitoring systems are generally easier to implement, but difficult to maintain, mainly due to people's natural reluctance to fill in forms, and frequently because they can see no outcome to the form filling process. Hence accident reporting is seen as a tedious task of no real benefit to the person charged with it. However, the task is rationalized if it is seen to lead somewhere.

Clearly, if risks do eventuate and accidents have happened, management needs to know why, so it can stop them recurring. Consequently, all reported incidents, whether notifiable or not, should be investigated and the following key issues examined:

What actually happened:

- to persons, plant and equipment
- when and where
- causes, direct and indirect
- effect, direct and indirect
- others involved or affected
- management response
- organization response
- adequacy of response?

What could have happened:

- worst case scenario
- why didn't worst case occur
- could it happen again?

The key aspects of a reactive monitoring system should include:

- the organization
- the person
- the task.

Results and analysis

In Chapter 1, the unreliability of data in accident statistics was noted and the relationship between direct and indirect losses discussed.

The whole point of measuring systems, monitoring and reporting, is to improve business performance. That means putting into place systems that work, are seen to work and provide useful reliable data to management, which can be used for the intended purpose. Accordingly, measurement and recording systems must:

- reliably and consistently collect useful data;
- be capable of analysis within the business process;
- record useful information.

Systems should be reviewed regularly for continuing relevance.

4.6 Auditing

Audit: 'any thorough going assessment or review'.

The performance of all systems, and of people, changes over time. It usually deteriorates, unless something is done to maintain it. Consequently, the purpose of auditing is two-fold: first, to maintain performance, and second, to ensure relevance and effectiveness.

Health and safety audits must be made by trained, competent people outside the department or activity being audited, and the following elements should be audited:

- policy
- organization
- planning and measuring
- reviewing.

Approach

There are many ways of conducting audits, involving both proprietary and bespoke systems, internal resources and external consultants, and various combinations.

Some organizations have developed scoring systems for the measurement of both quantity and quality of performance in those areas under review, but this is a development of the basic technique of auditing, which involves the following:

- **Vertical slice:** in which one **aspect** of policy, organization, planning and reviewing is examined.
- **Horizontal slice:** in which one **element** of policy, organization, planning and review is studied in detail.

In construction, a **vertical slice audit** could be applied to a contractor's employment policy, with particular reference to the use of direct labour or self-employed. About half of the workforce in construction is at least notionally self-employed and the usual reason for using the method is given as lower cost of employment in comparison to the total costs of direct employment.

Is this really the case? Has anyone within the organization studied the matter in recent years? Other questions which could be asked include:

Is direct labour more safe or less safe to use than self-employed?
Does the company have any records relating to either?

As an aside, and to the best of the writer's knowledge, there are no statistics available dealing with these questions and yet, certainly in construction, the answers are clearly important.

A **horizontal slice audit** could examine the organization and planning arrangements in place on the company's building sites, with particular reference to the way in which labour, direct and/or self-employed, is put to work.

Are they expected to turn up and get on with it?
What induction and task training are they given?
Are direct employees better trained, instructed and supervized than self-employed?

The audit could examine performance reviewing measures applied to labour in terms of quantity:

How many bricks per day are bricklayers expected to lay?
To what quality standards?
Against what performance standard?

For example, the British Standard relating to brickwork calls for it to be laid at the rate of four courses to every 305 mm of height, 10 mm horizontal and

85

vertical joints, of uniform colour and appearance using clean, new, undamaged bricks . . .

A point to note is that the business issues are indistinguishable from the health and safety issues.

Outcome of audit

The specific outcomes to be expected of a properly conducted audit are as follows:

- It should identify strengths and weaknesses within the company's health and safety management system, with particular reference to:
  - policy
  - organization
  - planning
  - measuring and reviewing.
- It should make clear recommendations for improvements in the areas of:
  - policy
  - culture
  - performance
  - reporting.
- It should provide a means of benchmarking the organization's performance against its competitors, in terms of **process** and **data**.
  - Process is a comparison of accident rates with comparable organizations and processes.
  - Data is an examination against current practices and techniques across all industries, or national averages.

Analysis and use of statistics

In conducting any audit or review process it is essential to understand the three basic analysis approaches:

Frequency rate

The frequency rate normally used is 100 000 hours to represent a person's average working life of 50 years at 40 hours per week, and the total number of man hours worked in the period, which may be a week, a month, a quarter or a year. Therefore, the frequency rate is calculated as follows:

$$\text{Frequency rate} = \frac{\text{Accidents} \times 100\,000}{\text{Total man hours worked in the period}}$$

For example, a firm has 16 accidents for a total of 200 000 man hours' work, therefore the frequency rate is calculated as follows:

$$\text{Frequency rate} = \frac{16 \times 100\,000}{200\,000}$$

$$= 8$$

When records are maintained over weekly, monthly, quarterly or annual periods, they rapidly reveal underlying trends which can indicate whether the firm's accident rate is rising or falling.

The formula can also be applied to different types of accidents. For example, serious, intermediate or minor. Work activities can be analysed. For example, excavation, brickwork and blockwork, carpentry and joinery which can be used for assessing performance and highlighting work activities where preventable loss is more likely to occur.

The frequency rate can also be used to compare departments, sites or companies.

Incidence rates

Incidence rates are usually calculated on the basis of the number of accidents over a given period per 1000 employees.

$$\text{Incidence rates} = \frac{\text{Accidents} \times 1000}{\text{Number of employees}}$$

For example, if there were 16 accidents and the firm employed an average of 100 employees, the formula would be:

$$\text{Incidence rate} = \frac{16 \times 1000}{160}$$

Frequency rate and incidence rate cannot be compared directly because they are based on different formula.

Mean duration

The mean durational average length of time lost per accident can also be calculated as follows:

$$\text{Mean duration} = \frac{\text{No. of man hours lost}}{\text{No. of lost time accidents}}$$

These formula are the starting point both in reporting systems based on compliance with RIDDOR 1996 and in any auditing and reviewing system.

4.7 Chapter summary and conclusion

Many people believe that health and safety management has much in common with the key components of good business practice, and that in many ways, the two are indistinguishable.

This chapter has examined both, and identifies the key components of successful health and safety management as follows:

- setting policy
- organizing staff
- planning and setting standards
- measuring performance
- audit and review.

In the business line, successful process management also has five key components, each of which is the broad equivalent of the above, and which include:

- modelling the process, or project
- choosing a process or project structure
- WBS/OBS/CBS systems
- an earned value mechanism for performance measurement
- a process review mechanism.

Organizations which manage quality in the business process also seem to have good track records in health and safety management. Construction has the second worst accident record in the UK. Its performance on quality is also very poor as three post-war enquiries have revealed.

When the essential pre-requisites for successful quality management are examined, it can be seen that the construction process, in its present form, has great difficulty in meeting most of them, and because of this, total quality management remains beyond the pale for most construction activity.

The major difference between construction and most other industries, in process terms, is the way in which designing, building and using have become separate and distinct activities in a manner that cannot be found in any other area of business in the UK.

5

Construction industry: current structure

5.1 Overview of the current structure

The origins of the modern construction industry can be traced to the Great Fire of London. Prior to that, clients constructed buildings made mainly from stone and timber and local materials immediately to hand. They would enter into 'bargains' usually with the master mason, who, in addition to carrying out his own work, would also control day-to-day operations on site and manage a series of other specialists, all paid directly by the client on a time and materials basis. The process would be administered by a further appointee of the client, whose distinguishing feature was not knowledge of construction, but numeracy and literacy. It was an age when most people could barely read or write.

The system relied heavily on close consultation between the client and the master mason, and the mason and all of the other trades. Basically, the client had to have a clear idea of what was required in terms of time and quality, and the mason was then responsible for achieving it. The system seems to have worked well for centuries.

More elaborate buildings, mostly commissioned by the Crown or the Church, involved architects and architecture, but usually in the form of a few exquisite hand drawings on parchment. It was still the master mason's task to turn the architect's ideas into practical reality on site.

The remnants of this system are still in use today, and a more sophisticated version of it can be found in the Construction Management Procurement Route.

Following the Great Fire, it was necessary to re-build London quickly, but much of the re-building was funded by insurers and they were not happy with the old system, which was basically cost plus reimbursement, a method that rarely appeals to clients.

So a system was developed involving the measurement of the completed works by 'separate measurers', and payment for it was made on the basis of what it was worth, as opposed to what the builder has spent on it. This practice is now enshrined in the legal notion of *quantum meruit*, that is that the person undertaking a piece of work should be paid that which is his worth within the specific circumstances of its use.

Over the next 300 odd years, a system developed whereby most clients wishing to procure construction, and there are many different types of client and construction, approached a consultant for advice before embarking on the process. This lead to the decline of the old way, lead by the master mason, and the development of a new process, lead by the newly developing professions, or groups of consultants differentiated by speciality. Hence the long march towards the separation of designing and building began.

The consultants concerned themselves with designing the works and administering the building contract and in so doing made sure that it was they that dealt directly with the client. They brought about the situation where specialist contractors simply built what consultants, not the client, now told them to build. The client was only the client of the consultants, he was the employer of the contractors.

Over time, separate disciplines developed, dealing with architecture, civil, structural, mechanical, electrical and public health engineering, surveying and so on.

Contractors, lead by Cubitts in London, increasingly gravitated towards general contracting, entering into a series of direct contracts with the specialists that in the old days would have been entered into by the client. This lead to the development, around the turn of the century, of the dynastic family firms, traces of which can still be seen today in Taylor Woodrow, McAlpine, Higgs & Hill, Laing, Cubitts and others.

Small companies tended to concentrate on the more complex specialist activities such as plumbing, heating and sanitary, electrical, structural steelwork and the like, and, whilst still working directly for clients, began to understand that their main source of work was through the new general contracting organizations.

Architecture established itself as the dominant profession, and for many years architects considered themselves to be the leaders of the professional team, and of the construction process.

Culture

From early Victorian times to around the 1970s, this was the dominant construction procurement mechanism, and its central features were that consultants were all employed on scale fees recommended by their professional institutions, whereas main contractors were nearly always appointed following a tender competition.

Whilst contractors were expected to manage all aspects of health and safety on site, there was no equivalent arrangement applied to the designing process.

Over time, an adversarial culture developed, rooted in two particular aspects of the system.

Completeness of design

The system contemplates a complete design prepared by consultants on behalf of the employer, and presented to the main contractor in sufficient time to enable him to construct, complete and maintain the works to the time, cost and quality standards set out in the contract, which includes design (see Figure 4.7).

In practice a complete design is rarely, if ever, prepared. Anything less than a complete design provided in sufficient time usually gives the main contractor the right to claim payment for extra time and money if he can show that his work has been delayed or disrupted because of incomplete or late design information.

Competitive tendering

Theoretically, contractors' tender prices should be based on the cost of building the complete design, plus a reasonable mark-up. In practice, the system has long since degenerated into a procedure to procure a price; the lowest price necessary to win the building contract. Clients, advised by consultants, have long recognized that this is the case, and have invited as many as 36 contractors to submit bids for the same project. Best practice would indicate that no more than six contractors should be invited. However,

**Table 5.1   Construction industry turnover (£ billions, 1990 prices) 1986–95.
Source: Department of Environment, Transport, and the Regions**

1986	1987	1988	1989	1990	1991	1992	1993	1994	1995
42.6	47.4	52.0	54.8	55.3	51.5	49.5	48.6	50.2	49.6

contractors have not helped themselves by actually bidding knowing they were competing against 35 other organizations.

Once into contract, the contractor has every incentive under the terms and conditions of contract to allege delay in receipt of design information, particularly if the client's needs change or he needs to instruct the consultants to vary the design. Consultants had no real incentive to challenge contractors' claims because they were paid a percentage of the final construction costs, which includes any claims payments made. However, as we moved from the gullible 1950s and '60s towards the litigious '70s and '80s, clients came to understand that late issue of design information could constitute professional negligence, which in turn, allowed clients to sue the responsible consultants and recover damages through their professional indemnity insurers, thus closing the loop of adversarialism and ensuring that it became a self-propagating process.

Simon (1944) and Banwell (1967) had both carried out major reviews of construction and had made recommendations for sweeping reforms, which had gone unheeded. So, by the late 1980s and early 1990s, the industry was in steep decline. The policies of successive governments had not helped. Table 5.1 shows the output of the construction industry between 1986 and 1995 (source: DETR: figures Billions: 1990 Prices).

These figures clearly show the cyclical, boom-bust nature of construction, and a look behind the figures is more revealing.

In 1989, the top 100 contractors and house builders generated over £2 billion of pre-tax profit, most of which was made in the housing or property development sectors. Between 1991 and 1993, the same 100 organizations lost over 1.2 billion. Some of them went bankrupt.

Among the factors influencing these events were:

■ changes in government policy regulating the public sector housing market;
■ changes in public sector procurement rules arising out of membership of the European Union;

- cyclical changes in the commercial property sector;
- a changing infrastructure market with particular emphasis on the newly privatized utilities;
- the introduction of the Private Finance Initiative (PFI);
- increased competition from American, Japanese and European contractors seeking a larger share of the UK market;
- fragmented nature of the industry itself;
- conflicting demands made on the industry by the range, disparity and widely differing needs of its client base.

The last two points are critical issues which are examined further below.

Fragmented nature of the industry

Effectively, the industry operates as a series of mini industries within, and contributing to, the process of construction. Each mini industry is differentiated by the vested interest groups it represents, and obviously, some vested interest groups have more power and influence over the construction process than others. The vested interest groups are:

Clients: who mostly use construction but are often involved in designing.

Consultants: who mostly design or provide cost advice.

Main contractors: who mainly build but also design and/or manage the process.

Specialist contractors: who mainly build but are increasingly assuming more and more design responsibilities and are the major employers of labour, both direct and self-employed.

Self-employed: who comprise about 50 per cent of the workforce, and without whom main and specialist contractors could not function. They have no official representation in any of the industry's quangos.

Managers: offering a specialist co-ordination role dealing with the combined activities of all other groups, essentially offering contracting services but on a consultancy basis.

Suppliers: who provide materials, plant and equipment either for sale or hire. They also play an important role in advising consultants on design issues and contractors on installation issues. They also undertake a significant research and development function.

Educators/trainers: who service all of the other groups by providing higher and further education through universities and colleges, and a range of further trades and professional development education and training through a number of public and private sector organizations.

Figure 5.1 Construction industry 1998: structure and organization

Seven of these vested interest groups (not the self-employed) are represented within the Construction Industry Board (see Figure 5.1) whose mission is to provide strategic leadership and guidance to the construction industry.

So, the fragmentation of the industry is there for all to see. Eight vested interest groups, 51 per cent of the workforce self-employed, over 200 000 contracting firms, only 12000 of which employ more than seven people. About 45 per cent of registered architects are sole practitioners (RIBA, 1993).

Typically, the industry turns over about £50 bn per annum, of which some 40 per cent by value is carried out by around 100 contracting organizations. The other 60 per cent is distributed between the remaining 199 900, of which about 20 000 per annum go bankrupt (all figures DOE unless otherwise stated).

5.2 Clients, their needs and their roles

In 1991/2 the public sector accounted for about 60 per cent of the industry's £50 bn output, with the remaining 40 per cent coming from the private sector. By 1995, the relationship had more or less reversed.

Within the public sector there are four major procuring bodies, Department of Environment, Transport and the Regions, Ministry of Defence, Scottish and Welsh Offices, plus a further 40 central government departments, 108 next-steps agencies and upwards of 400 NHS trusts (source: Levene Report) all of whom spend money on construction, a total of perhaps 600 odd clients.

Within the private sector, nobody knows how many clients there are, or who they are but they can range from BAA and Sainsbury, both of whom spend more on construction each year than most government departments, to home owners requiring minor domestic work. There must be, quite literally, hundreds of thousands of clients of the construction industry.

Masterman (1992) identifies five categories of clients which seems to cover most, if not all possible combinations (Figure 5.2)

Bennett and Flanagan (1983), Masterman (1992) and others describe clients' expectations and requirements of construction in the following way:

- certainty of time, cost and quality;
- no surprises during the procurement process;
- value for money;
- pleasing to look at;
- durable and easily maintained;
- reasonable running costs;

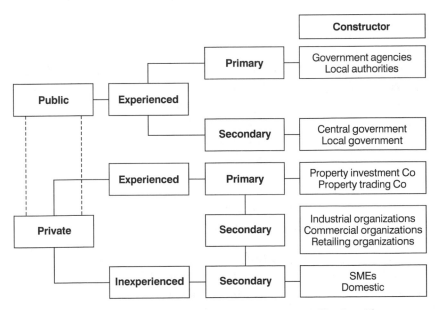

Figure 5.2 Categories of clients. *The constructor types are indicative. After a model by Masterman*

- no latent defects;
- good aftersales services;
- non-adversarial relationships with the supply chain;
- contractor to give worthwhile guarantees;
- early start on site and consistent progress;
- minimal interference from external sources;
- fit for purpose;
- customer delight;
- customer involvement and information.

Latham and clients

In 1993 Sir Michael Latham was appointed to undertake a joint government/ industry review of procurement and contractual arrangements in the UK construction industry.

In 1994 he published his findings in a report entitled 'Constructing the Team'. This is, without doubt, the most important and impressive examination of the industry in modern times and its recommendations have been implemented both by government and industry.

In his report, Sir Michael makes 76 recommendations under 30 main headings for sweeping reforms across the entire construction process, and he points out that implementation of his recommendations must begin with clients because they are at the core of the process. He also points out that clients are disbursed and vary greatly, and that they will commission projects which contribute to their wider business objectives.

Quoting evidence presented by Sir Bernard Rimmer of Slough Estates plc (Sir Bernard is a well known and respected figure in construction), Sir Michael points out that clients of the construction industry do not always get what they want. Rimmer's evidence compared the construction industry's performance to that of the modern motor car industry on nine crucial points (see Table 5.2; the author has added the tenth category, safety) and on each point, construction comes off worse.

As an aside, Rimmer's comparison of construction performance with the automotive industry's performance has become something of a liturgical chant since he made it. Everybody it seems, from major clients to distinguished academics and management gurus, constantly exhort construction to copy the motor car industry and the manufacture of its products.

Such criticism and advice is misguided and misplaced. Making buildings is not like making motor cars, and Rimmer (and Latham) were not suggesting that construction copies the manufacturing procedures of the motor industry, but instead, learns from its process and performance characteristics.

Table 5.2   Modern client satisfaction criteria for construction. Source: adapted from a presentation by Sir Bernard Rimmer

	Car	*Building*
Value for money	4	$3\frac{1}{2}$
Pleasing to look at	4	3
Free from faults	5	$1\frac{1}{2}$
Timely delivery	4	4
Fit for purpose	5	$2\frac{1}{2}$
Guarantee	5	1
Reasonable running costs	4	$2\frac{1}{2}$
Durability	4	2
Customer delight	5	2
Safety	5	1

The role of clients

In emphasizing the importance of the client's role in construction, Latham was reminding the industry of something it had always known, rather than telling it something new.

The Wood Report (1975), Bennett and Flanagan (1983), Turner (1990) and Masterman (1992) had all made significant contributions to better understanding of clients' needs and Latham's point was that the industry, and its clients, had failed to respond.

Some vested interest groups, mainly main and specialist contractors, had recognized and responded to changing client needs and requirements, by introducing new procurement routes intended to give clients alternatives to the discredited traditional system. However, their efforts were strongly opposed by other vested interest groups, mainly consultants, because the new routes presented a threat to their perceived domination of the construction industry.

5.3 Procurement routes available to clients

When Latham published his report in 1994, six main procurement routes were available and they are:

- traditional;
- design and build;
- design and manage;
- management contracting;
- construction management;
- engineer, procure, construct.

The major differences between each of the above routes are as follows:

- Each distributes business risk along the supply chain in a different way.
- Each deals with the integration of designing with constructing in a different way.
- Each requires different levels of involvement of clients.
- Each requires the vested interest groups to modify their behaviour and the services they provide.
- Each requires a different and complex contract strategy to implement it fully.

Clearly, the choice of procurement route plays a vital part in meeting the client's needs and expectations, and in delivering a successful outcome to any construction project.

Procurement assessment criteria (PAC)	Procurement arrangements option (PAO)					
	Design and build Balance of risk		Traditional Balance of risk		Management Balance of risk	
	Client	Contractor	Client	Contractor	Client	Contractor
1. Programme (timing)	1	4	2½	2½	1	4
2. Variation	1	4	2½	2½	4	1
3. Complexity	1	4	4	1	4	1
4. Product (quality)	3	2	1	4	1	4
5. Price	1	4	1	4	4	1
6. Competition	1	4	2½	2½	4	1
7. Responsibility	1	4	2½	2½	2	3
8. Risk (overall)	1	4	2	3	4	1
9. Safety	2	3	3	2	3	2
Scale: units of 1 to 5						

Figure 5.3 Risk assessment criteria between procurement routes. Source: adapted from Turner, 1990

Latham repeated the point made by others, that procurement choices should not revert to the default position of the traditional route. Instead, a careful risk assessment should be made prior to choosing a route, based upon eight procurement assessment criteria (PAC; the author has added the ninth criterion, safety – see Figures 5.3 and 5.4).

The choice of route determines the contract strategy to be applied in assembling the supply chain needed to deliver the project to the procurement assessment criteria established by the client at the outcome.

The choice of route automatically determines one further issue of vital importance to clients, namely, the health and safety management strategy to be applied, because it configures the optimum CDM compliance arrangements within the project.

CDM compliance within construction projects

The reason the procurement route determines the health and safety management strategy is that the statutory appointments seem to fall naturally into place in a particular way within each procurement route.

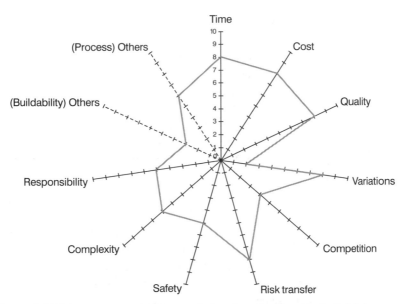

Figure 5.4 Risk measurement. This graph describes the 'quantum' of risk in a project, measured against the project-specific importance placed upon each head of PAC by the client. If his priorities change, so does the importance he places on the PAC, so the shape of the graph will change.

However, it is important to note that the seemingly natural appointments are by no means the only ones that can be made. In practice, the regulations are written in such a way as to allow the broadest possible range of choices of statutory dutyholders, subject always to the over-riding considerations of competence and adequacy of resource.

Two appointments do fall naturally into place without choice. Dutyholders are either designers, and contractors, or they are not.

Three appointments can be the subject of choice. Clients can appoint a client's agent, and the planning supervisor can be anyone. So can the principal contractor, provided he is a contractor. All of these appointments are subject to review and can be changed at any time. Indeed, if there is evidence that competence is less than satisfactory, then the appointments must be changed.

Types of construction

In the UK, construction falls into the following broad categories:

- housing and housing maintenance;
- general building and building maintenance;
- civil engineering, and associated maintenance;
- process engineering and associated maintenance.

This is particularly relevant because a series of standard forms of contract have been developed for use within each category, so the contract matrix used to implement a building project would probably not be the same as for a civil engineering project.

These issues are discussed further in a later chapter.

5.4 The traditional route

Contract arrangements (see Figure 5.5)

In a traditionally procured job, the employer appoints a team of consultants to prepare a complete design which will normally include drawings, design detail, specification and sometimes a bill of quantities. The design is used to procure competitive tenders normally from six suitably qualified and experienced main contractors.

The contractor submitting the lowest price is normally appointed, and his obligation is to construct, complete and maintain the works to the time, cost and quality standards set out in the contract. The contract normally comprises

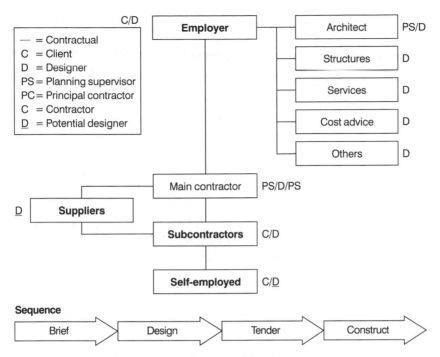

Figure 5.5 Contractual/CDM compliance: traditional route

the contractor's offer, the client's acceptance and the design prepared by consultants.

In a building contract, the JCT standard forms of contract would probably be used to implement the contract matrix, whereas in a civil engineering contract, the ICE terms and conditions would probably be used. However, a contract known as the ECC (formerly NEC) could be used for either.

CDM-compliance arrangements

In a fully designed job, the CDM-compliance arrangements falling naturally into place are as follows.

Client: The employer is usually best placed to undertake this role.

Designers: All of the consultants are designers.

Principal contractor: The main contractor is best placed to be the principal contractor. Because certain items of temporary works are also 'design', namely formwork, falsework and scaffolding (see CDM Guidance Note 71), he will also be a designer.

101

Contractors: All subcontractors and some self employed persons are contractors, and formwork, falsework and scaffolding sub-contractors are also designers.

Planning supervisor: There are two 'natural' arrangements. One of the consultants undertakes the role for the duration of the project, or, one of the consultants undertakes the role for the design phase and the main contractor takes over during the construction phase.

Alternative arrangements

Clients agents: A client's agent can be appointed. This is appropriate where, either the client is inexperienced, or where the identity of the client may not be clear; for example, in a multi-funded project, or where a holding company is being used as a development vehicle.

Planning supervisor: In many cases, there is a strong case to be made for the use of an independent planning supervisor. This is particularly so where there are complex health and safety issues likely to arise within the project design, or where phased or sectional completion is required, or where a complex design is envisaged, which will require exceptional co-ordination measures. The planning supervisor function provides a superb platform from which to obtain a continuous overview of designers and design activity.

Principal contractor: One of the subcontractors could be the principal contractor, or the role could be made the subject of a separate work package and appointment.

These alternatives do not fall naturally into place, but they are valid. There are also other arrangements which could be made, always subject to satisfactory competence and adequate resource allocation.

Benefits and pitfalls

The claimed benefits of the traditional system are as follows:

- It is well known, tried and tested.
- If design is complete at tender stage, price certainly is achievable with returned tenders.

- Design can be completed without time pressure from contractors, and the possibility of delays for late receipt of information.
- The completed design ensures a common basis for tendering.
- Different tendering methods can be employed to gain the benefits of different type of competition.
- Contractor design ideas can be taken into account and can be incorporated into the contract arrangements to obtain the benefits of acceleration of programme time on site.

Some common pitfalls are:

- Designing and building is separated and this tends to create vested interests within the project team. Inevitably, this leads to adversarial attitudes and conflict between consultants and contractors, impacting on the client.
- The system takes longer than most other routes.
- If variants are used, uncertainty is introduced and contractual responsibility diluted.

However, the major problem is prosaic. In practice, complete designs are never prepared and consultants have in recent years increasingly turned to contractor design, which is written in the main contractor's tender and contract. It is not unusual for up to 40 per cent of a fully designed job to be *de facto* contractor design.

This immediately raises questions of CDM compliance. Specifically, it means that the main contractor and many of the subcontractors and perhaps some suppliers, and even the self-employed, will be carrying out design within the CDM definition. So, instead of perhaps five or six consultant designers, there may now be 50 or 60 contractors who also have design duties under the regulations.

This is turn impacts on those appointing those designers. The person making the appointment must ensure that the appointees are competent and adequately resourced, and the planning supervisor must ensure that they all co-operate with each other. In this way, the health and safety management structure and contract arrangements have been changed by the contractual arrangements made.

The regulations were made to embrace exactly this kind of situation, and to ensure that appropriate health and safety management arrangements are made. In the past, experience has shown that consultants rarely fully consider the health and safety implications of their design decisions, and contractors who are designing, tend to apply lower standards than those which would be applied by consultants.

That is why the regulations set performance standards which must be met, but if the default position occurs, that is an inappropriate health and safety management arrangement is made, or no such arrangement is made at all, then there will be a clear audit trial established, which might lead to successful criminal prosecutions.

If the modified traditional route is used, involving contractor design, the point to remember is that contractors will be 'designers' and this will impact on other designers, the planning supervisor and the principal contractor – and perhaps also the client. It is essential that all CDM-compliant work is design risk assessed before any work is put in hand on site and there must be clear procedures in place to ensure that this happens.

5.5 Design and build route

Contractual arrangements (see Figure 5.6)

In a design and build contract, the employer prepares the 'employer's requirements' which set out in terms of function and performance, the standards which the building should achieve. Sometimes, it requires the input of consultants to assist the employer in the preparation of this document, which is then put out to competitive tender, again, to perhaps three or four suitably qualified and experienced design and build contractors. The tendering exercise has two broad objectives:

- to obtain innovative design solutions;
- to procure comparable tender prices.

Contractors respond with the 'contractor's proposals' which set out in detail the way in which they propose to design and build the works to the time, cost and quality standards set out in the employer's requirements.

The contract comprises tender offer, employer acceptance, employer's requirements and contractor's proposals. Consequently, any errors or omissions or weaknesses in the former simply flow directly into the latter.

Different families of standard forms of contract and subcontract are available for use in building, civil engineering and process engineering.

CDM-compliance arrangements

In a design and build job, the CDM-compliance arrangements falling naturally into place are:

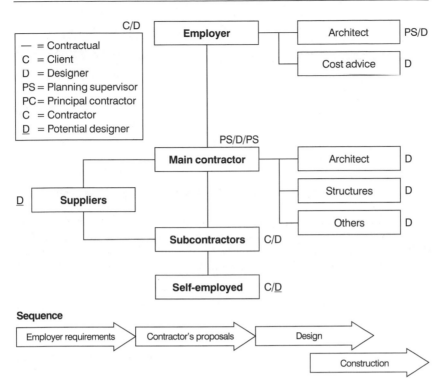

Figure 5.6 Contractual/CDM compliance: design & build and design & manage

Client: The employer is usually best placed to undertake the role.

Designers: The employer's requirements are design, so whoever prepares them is a designer. This will usually be the employer, and any consultants he uses. The design and build contractor will also be a designer and will nearly always use a combination of consultants, specialist contractors, suppliers and the self-employed to assist in preparing the design. They will all be designers.

Principal contractor: Usually the design and build contractor is best placed to undertake this role.

Contractors: All of the design and build contractors, subcontractors and self-employed will be contractors. Sometimes the employer may use direct contractors of his choosing to work on the project at the same time as the main contractor and his team. Figure 5.7 illustrates this point. The direct contractors are in contract with the employer, but statutory responsibility for their performance rests with the principal contractor.

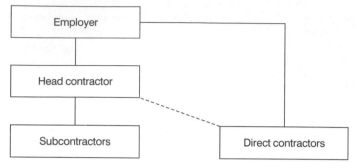

Figure 5.7 Direct contractor arrangement This arrangement can occur in all routes. The employer may wish to use directly employed contractors on the project, and the head contractor will probably allow him to do so. The head contractor is not in contract with the direct contractors, and therefore has no powers of instruction. The CDM Regulations gives two statutory powers to the principal contractor, sufficient to enable him to achieve his own compliance.

Planning supervisor: Either the client or one of his consultants can act as the planning supervisor up to the preparation of the employer's requirements. It is then sensible to appoint the design and build contractor once a choice has been made and he then remains in place until the design and build contract is finished.

Alternative arrangements

The obvious alternatives can include:

- A client's agent appointment, for the same reasons as before.
- An independent planning supervisor is particularly appropriate for this route, where essentially the employer hands over control of the design to the main contractor at an early stage. An independent planning supervisor can provide a useful overview of the way in which the design of the project is progressing.

Benefits and pitfalls

Design and build is best suited to projects where function and performance is more important than prestige and appearance. The claimed benefits of the system include:

- Single-point responsibility is obtained from one organization.
- The employer can obtain an early indication of the total cost commitment to be entered into.

- Designing and building are better integrated, and this produces savings through improved buildability and innovative design solutions.
- Construction including the design and operation of processes (chemical petrochemical, utilities and pharmaceutical) can achieve much greater time savings by overlapping designing and building.

The pitfalls can include:

- Making meaningful comparisons between alternative design and build tenders can be difficult for inexperienced employers.
- The cost and risk of employer's changes can be much greater once work starts on site.
- Tender cost falling on contractors tends to inhibit them, and so they offer *de minimus* bids based on partially developed designs. This can be overcome by the use of two-stage tendering.

In terms of CDM compliance making the initial arrangements and appointments is usually fairly straightforward. On the employer's part, the important thing is to recognize that the employer's requirements are 'design' and to ensure that the hierarchy of risk control is applied. An outline health and safety plan must also be included among tender documentation sent to tendering main contractors, and there must also be a health and safety file available.

Within the supply chain, the design and build contractor's role is of course crucial, for they are principal contractor, planning supervisor and designer, if the natural configuration is used. Very often the main contractor will expect subcontractors, and sometimes suppliers to provide all of their own design detail requirements, and they will simply not comply unless managed.

At site level, workers are usually employed on a lump sum price basis, and in the absence of adequate information on where to put a pipe run, or a radiator, or a handrail or a power socket, they will simply put these things where they think they ought to be.

Again, the objective of the CDM regulations is to stop this from happening by co-ordinating the design in advance of work being put in hand on site, but the default position is a clear audit trail leading to criminal prosecutions if this does not happen.

5.6 Design and manage route

Contract arrangements (see also Figure 5.6)

Design and manage is a variant of design and build, wherein the main contractor does no work on site, but simply provides a management framework and resources. Where a formal contractual strategy is implemented, then the arrangements both in contract and in terms of CDM compliance are indistinguishable from design and build, and for that reason it is not widely used for it offers no measurable advantages, and suffers from the disadvantage that it usually costs more to implement because the main contractor has no resources of their own on site.

CDM-compliance arrangements

The appointments falling naturally into place are the same as design and build, and the same alternative compliance arrangements are also relevant.

Benefits and pitfalls are similar, but there can be one major area of concern in this route, where it is not formally chosen and implemented using a recognized contract framework.

In many organizations which procure construction, design and manage tends to be a default position. This is particularly so in maintenance or facilities management, in regional or local government, housing associations, NHS trusts, Ministry of Defence establishments, universities and large private sector commercial organizations. Frequently, these organizations employ property services, facilities managements, technical services or some other central procuring mechanism where contractors and specialists are chosen either at random, through a select tendering list, or from a maintained list of contractors.

For example, an NHS trust may decide to refurbish a ward, and might appoint:

- a painting and decorating contractor;
- a floor covering contractor;
- a suspending ceiling contractor;
- an electrical contractor;
- a mechanical contractor;
- an IT contractor.

The trust's estate management department will probably package the work into self-contained contracts where each specialist provides all of its own attendances, and makes its own interface and co-ordination arrangements with the other specialists. The object of this exercise, and way of procuring, is to do away with the main contractor's services. The main contractor is seen as an unnecessary expense, as in the trust's eyes, it plays no visible role in terms of work done on site. The employees of the trust can often regard the main contractor as a nuisance, and much prefer to deal directly with the specialist contractors, subcontractors and suppliers, and it has to be said, the reverse is also true. Everybody wants to talk to the client.

Effectively, the trust takes the view that it will provide all of the design and management activities that it thinks are necessary, and act as principal contractor itself (*CDM Reg. 6 makes it clear that to be a principal contractor, an organisation must itself be a contractor*) then expects the specialists to get on with it themselves. What the trust then does is to administer a series of specialist contractors.

What it does not do is manage a construction process, and that is precisely what a principal contractor is appointed to do, at least in respect of health and safety, but as has previously been shown, these cannot be separated from the broad process.

Employers who procure construction in this way deserve to go to jail, because the inevitable result is that everybody involved falls into breach of both the CDM Regulations and the Construction (Health, Safety and Welfare) Regulations. Contracts procured in this way always seem to be dirty, untidy and characterized by inefficient working, visible waste and poor quality work. It is unfortunate that there are no statistics available dealing particularly with the safety records of these types of contracts, but experienced observers know full well that they are perhaps the most unsafe in the entire industry.

Operatives are expected to have their meal breaks in their vehicles, or in some half finished plant room. There are usually no lavatories, changing or washing accommodation provided and education and training of the workforce is non-existent. Information delivery and dissemination is done in haphazard fashion, if at all, and they are characterized by complacency at site level and adversarial attitudes between participating organizations.

The writer was once asked to rescue a project similar to the arrangement described above which had gone badly wrong. It was a 26 week refurbishment project at a defence works establishment which had fallen 20 weeks behind programme and was about to incur a massive cost overrun when the employer decided to bring in a main contractor. After an initial appraisal, work was suspended on site and the employer's designers were instructed to complete the design information needed by the package contractors. They had previously issued 147 architects instructions containing over 600 variations and each and every one had been endorsed with the words 'This instruction is clarification of existing design details and carries with it no entitlement to further time or cost.'

Designers were taken off all other duties not connected with this project and they were forbidden to use faxes, handwritten messages, verbal instruction or letters. They were required to produce relevant, properly detailed drawings and specifications. It took them nine weeks to do so whereupon the design was frozen and no further development of it was permitted. Only when a complete design was available was the main contractor allowed to start work. Contract works on site where then completed within five weeks.

In order to avoid falling into this trap, employers have perhaps three realistic choices:

■ Act as principal contractor themselves, provided they are competent to manage the process.
■ Use a main contractor.
■ Appoint one of the specialist subcontractors.

The default position, that is no main contractor and therefore no principal contractor, is illegal.

5.7 Management contracting

Contractual arrangement (see Figure 5.8)

In a management contracting arrangement, the employer appoints a team of consultants to prepare feasibility concept and outline designs.

After a tender competition, a management contractor is brought on board at an early date, usually on similar terms and conditions to the consultants. The prime function of the management contractor at this stage is to assist the

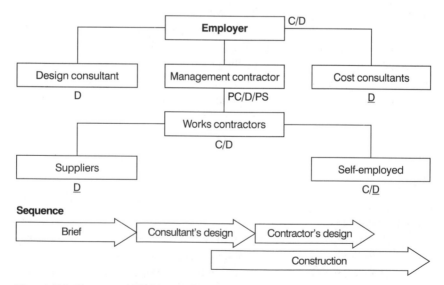

Figure 5.8 Contractual/CDM compliance: management contracting

110

consultants to develop a series of integrated works packages and to improve the amount of innovation and buildability contained in the design.

Works packages are then put out to competitive tender and the management contractor enters into a series of contracts with them to complete and detail the remaining design and to construct, complete and maintain the works contained in the packages to the time, cost and quality standards established by the employer.

CDM-compliance arrangements

In a management contracting project, the CDM-compliance arrangements falling naturally into place are as follows:

Client: The employer is usually best placed to undertake the role.

Designers: Consultants, the management contractor and works contractors are all designers, and some suppliers and self-employed can also have design duties.

Principal contractor: Usually the management contractor is best placed to undertake the role.

Contractors: All of the works contractors and any subcontractors or self-employed people that they use are contractors. Also, any direct contractors employed on the site during the currency of the management contract will be contractors under the control of the principal contractor.

Planning supervisor: Most management contractor appointments are made early, at the same time or shortly after the consultants and before any significant amount of designing has taken place. Consequently, it makes sense for the management contractor to undertake the role throughout the project.

Alternative arrangements

A client's agent and an independent planning supervisor are two obvious alternatives, and there is much to recommend both.

Management contracting is best suited to large, complex projects – often fast track commercial developments where time is of the essence and ownership of the project may not be clear.

A client's agent appointment brings clarity to the client's role and ensures consistency in the discharge of the client's duties.

The independent planning supervisor introduces method and rigour into the design process.

Benefits and pitfalls

The benefits claimed for management contracting include:

- It enables design and construction to proceed in parallel and can result in better co-ordination and co-operation between overlapping packages.
- Price competition between trades packages is obtained.
- Adversarial attitudes are seen to be inappropriate because contract terms do not incentivize adversarial behaviour.
- Programme time can be reduced significantly.
- The employer can play an active part in the whole process.

Some of the pitfalls can include:

- Consultants, who often feel that they are not given enough time to develop designs or to exercise design skills, become dissatisfied.
- Design quality is often compromised with aesthetics suffering at the expense of function and performance.

Management contracting is generally used on large city-centre building projects, usually distinguished by complexity and time. There are a large number of general contractors offering management contracting services, but only about six to ten companies in the UK perform the service well. When it is done badly, it defaults to a sort of 'post box' operation with the management contractor standing between the works contractors on the one side and the employer and consultants on the other. In terms of health and safety management, provided the naturally occurring arrangement is used, responsibilities are clear. If difficulties arise, they usually do so from complexity and time constraints.

Complexity occurs as a result of split design responsibility. For example, a consultant might design 'rainscreen cladding' which might involve structural steelwork, extruded aluminium subframes, metal and glass panels and insulation. The consultant will draw pictures of what he wants the cladding to look like and he will also write specifications for its technical and functional performance. Works contractors will then be expected to complete and detail this design.

Overall, this might require the combined input of the perhaps two or three consultants and six or seven contractors. Multiply this by 50 or 60 works package contractors, and it is easy to see how major co-ordination problems can occur. The planning supervisor must ensure that all designers co-operate with each other.

Time is affected because the route is marketed on the basis of the main contractor's ability to actually optimize these works packages, and co-ordinate their activities in less time than in other routes whilst overlapping designing with building. The inevitable outcome is that short cuts are taken and work starts on site with incomplete or no design, and operatives are forced to decide, as work proceeds, matters that should have been dealt with further up the supply chain.

Services installations are particularly prone to this problem, partly because they are complex, partly because they are usually put into the building quite late in the programme and partly because it takes so many different people to design them. Particular problems occur with:

- horizontal and vertical distribution through the building;
- availability of space within the building;
- concealment of components;
- access for cleaning, maintenance and replacement.

People begin to take foolish risks in terms of knocking holes through walls or floors, drilling holes with unsuitable equipment, working off ladders or the ubiquitous plastic milk bottle crates, when they should be using proper scaffolds and platforms. They also do the same in terms of deciding, in the absence of any other information, where to put hot and cold water pipes, cable runs, steam pipes, fire alarms and the like. This results not only in unsafe working on site but can also result in buildings that are unsafe to clean, maintain, alter, refurbish, operate and ultimately demolish.

The route requires two essential elements:

- an active informed employer, and
- a proper management contractor.

5.8 Construction management

Contractual arrangements (see Figure 5.9)

In a construction management arrangement, the employer enters into a series of direct contracts with consultants to prepare feasibility concept and outline design. The construction manager is appointed at the same time as and on the same basis as the other consultants and his job is to work with them to identify appropriate trade packages.

113

Trade packages are then usually put out to competitive tender, and the employer then enters into a series of direct contracts with trade contractors to complete the design started by consultants, provide all necessary design detail and build, operate and maintain the work described in the package to the time, cost and quality standards set out in the bidding information.

CDM-compliance arrangements

In a construction management project, the CDM arrangements falling naturally into place are:

Client:	The employer is usually best placed to undertake the client's role.
Designers:	Consultants, the employer, the construction manager and the trades contractors will all be designers, along with some of the suppliers and subcontractors and the self-employed.
Principal contractor:	This is usually the construction manager.
Contractors:	All of the trades contractors, their subcontractors and the self-employed will be contractors.
Planning supervisor:	The construction manager is best suited to the role because of his involvement with all of the consultants and trade contractors throughout the project.

Alternative arrangements

Again, a client's agent could be appropriate. An independent planning supervisor appointment can bring benefits, but tends to be a duplication of the construction manager's role.

A separate works package could be made of the principal contractor role, but again this tends to duplicate many of the construction manager's functions.

Benefits and pitfalls

Construction management exponents, despite the obvious similarities, complain strongly when the method is compared with management contracting. Construction managers say that it is a philosophy rather than a formal procurement route. There are perhaps two or three exceptional construction

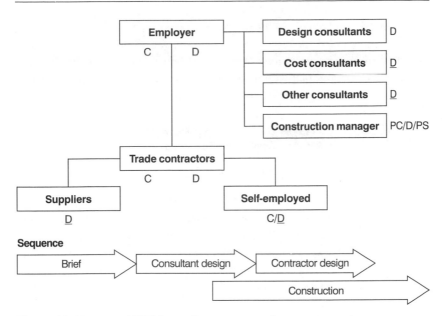

Figure 5.9 Contractual/CDM compliance: construction management

managers in the UK, although the system is in much wider use in the USA and Japan. In essence, it is a reversion to the old way of procuring construction prior to the Great Fire of London, but obviously with much more rigid time, cost and quality controls. So far in the UK, its use has been limited to a relatively small and specialized market sector, mainly high-quality large complex building projects and the central London office market, and its track record has been excellent in term of customer satisfaction. It really does appear to fulfil the claims made of it by construction managers, but there are some notable underlying reasons for this.

- It is generally used by experienced, intelligent clients who have expert insight into the construction process, and this is a vitally important factor.
- Construction managers themselves are highly skilled, competent organizations who apply a total quality management approach to the process. They employ the best people and train them thoroughly.
- Great care and attention is paid to the selection and appointment of consultants and contractors. They are selected as much for their ability to contribute to the construction management approach, as for their overall technical skills.
- The route seems to work because people try harder and behave better than in other procurement routes.

115

The benefits and pitfalls of the route are similar to management contracting, but they are avoided, or have been to date, by the TQM philosophy which underpins the route. Clients are exposed to more risk from the method because they are in direct contract with trades package contractors, which is why it is of vital importance that clients are experienced.

5.9 Engineer, procure, construct (turnkey)

Contract arrangements (see Figure 5.10)

The abbreviation applied to engineer, procure, construct is 'turnkey', because the employer's only involvement is said to be to turn the key in the front door on the day he wants to take possession of the building. The reality is somewhat different and requires a certain amount of imagination stretching.

There are two approaches in common use. Firstly, the employer describes the service or process to be purchased from suppliers, and invites bids from them along the lines:

'I wish to purchase ten tonnes of steam and ten megawatts of power each day for the next ten years. At the end of that time, I will rebid the service

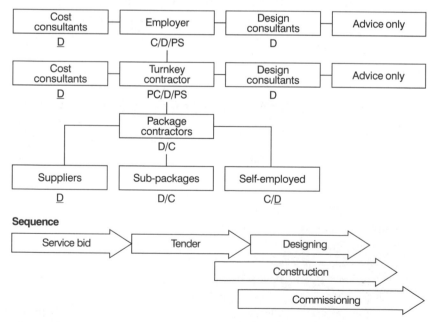

Figure 5.10 Contractual/CDM compliance: engineer, procure, construct (turnkey)

provision and guarantee to include you on my list of people invited to tender'.

The employer, and the bidders, understand from this that what is really required is a combined heat and power plant, for that is the only way of meeting the service need from a single energy source. So, contractors will bid to design, build, operate and maintain (DBOM) a combined heat and power plant for ten years, knowing they will almost certainly win the contract for the next ten years as well.

Second, the employer actually prepares on outline design similar to the employer's requirements in the design and build route, but included within the outline design is a **process**. The method is in common use in the oil industry where, for example, a tender enquiry might say:

> Design, build, operate and transfer (DBOT) a new crude distillation unit complying with the ACME Oil Company standard design type 2, copy enclosed, with a process capacity of 5000 tonnes of crude oil per day. Process capacity is to be in accordance with ACME process standard number 6.

Here, the employer, ACME Oil, has developed a series of standard designs for crude distillation units and various types of crude distillation processes because he quite rightly considers himself expert in the business of owning and operating oil refineries and he knows exactly what is required. However, he does not wish to bear the business risk associated with building a new crude distillation unit and he wishes to pass that risk to the supply chain.

The game that is played is that he has **not** prepared any design as far as the contract documents are concerned, and to the extent that tendering main contractors choose to use standard design type 2 and process standard number 6, then it becomes their design and they take full contractual responsibility for it.

In both cases, the employer may enter into separate agreements with consultants to assist him to prepare either his service description or the equivalent of the employer's requirements. However, there is then one contract between the employer and the turnkey contractor, followed by a matrix of supplier/subcontractor contracts between the contractor and the supply chain.

Turnkey is used extensively in procuring process-related construction, particularly in the water treatment and power generation, transmission and distribution sectors using standard forms of contract of an engineering nature. A contract called FIDIC is in common use.

117

CDM-compliance arrangements

The CDM compliance arrangements falling naturally into place are:

Client: The employer is best placed to undertake the client's role.

Designers: The client, any consultants he uses, the turnkey contractor, his consultants subcontractors and suppliers and their subcontractors and the self-employed can all be designers.

Principal contractor: The turnkey contractor is best suited to be the principal contractor.

Contractors: Everyone working on the site below the turnkey contractor in the supply chain is a contractor.

Planning supervisor: The turnkey contractor is best suited to be the planning supervisor.

Alternative arrangements

Once more, a client's agent and independent planning supervisor appointment strategy is worth serious consideration, and in some cases, can be the natural compliance arrangement.

Projects in the petrochemical and utilities sectors can be very large and complex and can involve multi user, multi process sites, separated from each other and spread over quite large distances.

For example, a new CHP plant might be used for generating steam and power at one site, transmitting it to several others, and distributing it around those sites, which may even be occupied by different organizations. The identity of the client may not be clear at all in that scenario, and there is an obvious case for a client's agent.

Similarly, there may be several different types of construction including building, civil engineering and process engineering, involved within one project and some complex occupational health and safety issues may occur. Clearly, there is a strong case for a properly resourced external planning supervisor organization.

Benefits and pitfalls

From the employer's point of view, the objective of the route is to pass all business risk to the supply chain. It is rarely possible to do that, but employers and their lawyers are never deterred from trying, hence some extremely onerous terms and conditions of contract are used, or drafted, in this route.

Naturally, employers seek to place the business risk with the main contractor, who uses equally onerous contracts to distribute the business risk along the supply chain. Some people actually believe that when diversified in this way, risk goes away.

The benefits of the route are similar to the management routes, that is:

■ designing and building are unified;
■ single point responsibility is achieved;
■ designing and building can be overlapped;
■ significant time and cost savings can be achieved from the overlapping;
■ innovative design solutions improve buildability;
■ better buildability leads to better quality.

In addition, the route is widely known and accepted internationally. This has two benefits. First, many overseas companies have bought into the UK infrastructure since privatization. They use the method themselves and have brought it with them. Second, because UK companies are now acquiring more and more experience of the method, they are becoming much more competitive overseas themselves.

A third benefit relates to the PFI. So far, PFI has failed in the UK because it is perceived as too complex and costly. Eventually, the government must either abandon it or make it work, in which case, turnkey contracting is, for the most part, the natural way of procuring any construction needed within the service provision contract.

One major pitfall of the route is the incompatibility of the business risk passing philosophy with health and safety risk management philosophy; that is, risk lies where it falls. Sometimes, employers and their advisers will write contracts that attempt to pass health and safety risks falling on the client, to others in the supply chain. It is important to remember that the client's duties must be carried out by the client, or client's agent, and whilst the actions in the duties can be delegated, responsibility cannot be passed; it rests where it falls.

Another pitfall is that 'contracts' are sometimes confused with 'projects'. Imagine our CHP plant is built at site A and is to transmit steam and power via overhead line B to site C, and overhead line D to site E. At sites C and E, there are separate termination structures F and G, from whence steam and power is distributed around both sites, via distribution schemes H and J. A to H are all let on separate contracts, but have the same employer and the same turnkey contractor.

What do we have here, one project? A new CHP plant and transmission and distribution system covering five sites and including eight contracts (A to H), or eight separate projects, consisting of eight individual contracts, A to H?

Many people lean towards the latter solution for the same reasons that lead them to believe that business risk can be diverted out of the supply chain. However, health and safety law doesn't work that way.

If the latter view is taken, eight separate projects, then eight separate CDM-compliance arrangements must be made, that is eight planning supervisors, eight principal contractors, eight health and safety plans, eight health and safety files and eight sets of procedures. There might be work activity occurring on contract H, perhaps five miles away, which has a significant impact at site A. Operation and co-ordination arrangements then become a nightmare and in the writer's view this configuration defeats the single-point responsibility philosophy which underpins CDM. At a more practical level, it would either cost a great deal of money to implement, or it would simply not work.

In most cases, the CHP plant and all of the other seven pieces of work are one project comprising eight separate contracts which may or may not be carried out in sections or phases and CDM-compliance arrangements must be made project wide.

Frequently, this is not what very senior management wishes to hear, and it will overrule the advice of health and safety line managers, simply because the one project scenario opposes their business risk management philosophy. This is one area of CDM where test cases would be useful to bring clarity to what does appear to be a confusing situation.

5.10 Chapter summary and conclusion

For millennia, there was only one way of procuring construction in the UK. However, from around the time of the Great Fire of London, things began to change, largely because of the way the vested interest groups contributing to the construction process responded to client needs and demands.

The changes can be seen in two distinct and separate phases. In the first, from around 1666 to the 1960s, the traditional procurement system was developed, featuring the separation of designing from building, and establishing strong leading roles for consultants within the process.

The second stage, from the mid-1960s to the present, saw a contractor-led drive towards co-operative systems seeking to unify design and construction, challenging the established role of consultants and offering management solutions as an alternative to the discredited traditional system.

Consequently, clients now have a choice of six major procurement routes, each one implemented by the use of complex contract matrices. The intent and purpose of the route is to give clients a range of choice in the way in which business risk is distributed and managed, and who does the managing.

Consequently, the choice of procurement route and contract strategy has a major impact on the health and safety management arrangements which are made in respect of all CDM compliant projects. The choice of procurement route configures the supply chain used to deliver the project, and from there, certain CDM compliance arrangements seem to fall naturally into place, subject always to satisfactory competence and adequate resources.

This chapter has examined the development of the six procurement routes, the principal characteristics of each, the naturally occurring CDM compliance strategies, the alternatives and the benefits and pitfalls of both the routes themselves and the health and safety management arrangements that flow out of them.

Risk assessment of both design and construction is central to any health and safety management arrangements made, within the chosen procurement route. The main features of risk assessment in construction are examined in Chapter 6.

6

Risk assessment

6.1 Risk assessment and construction

To those not familiar with its ways, the construction process can be baffling. Even to those who are familiar with it, indeed to those who earn their living from it, things are far from clear. In construction the process of designing is separated from the process of building and constructing in a way that cannot be found in any other industry, and it is this separation that goes to the root of construction's difficulties.

In all important aspects, designing and building have become two separate processes. Since 1995 they have been regulated by two separate sets of risk assessment requirements.

Designing and managing is regulated, *inter alia*, by the CDM Regulations, and design risk assessment must be carried out in accordance with **Regulation 13: Requirements on Designer**.

Building and constructing is regulated by many different sets of regulations, but the way in which building work is to be carried out must be risk assessed in accordance with **Management of Health and Safety at Work Regulation 3: Risk Assessment**.

The approach to risk assessment remains refreshingly simple.

At CDM Regulation 13, designers are required to (*inter alia*):

- identify hazards in their designs;
- identify risks arising from hazards;
- eliminate, reduce, or control the risks they have created.

At Management of Health and Safety at Work Regulation 3, employers (that is, contractors) are required to examine their work methods and:

- identify hazards in work methods;
- identify risks arising from hazards;
- eliminate, reduce, or control the risks;
- design and implement a safe work system;
- monitor, audit and review the system.

The Construction Industry Training Board (CITB) is a statutory body established in 1964. Its main functions are:

- to develop areas of training of vital importance to the future of the construction industry;
- to ensure there is an adequate supply of people trained to appropriate standards to meet the industry's needs.

The CITB defines hazard and risk in the following way:

Hazard: The potential to cause harm, including ill health and injury; damage to property, plant, products or to the environment; production losses or increased liabilities.

Risk: The likelihood that a specified undesired event will occur due to the realization of a hazard by, or during, work activities or by the products and services created by work activities.

The statutory definition of hazard and risk is made at Regulation 3 of the Management Regulations:

Hazard: A hazard is something with the potential to cause harm (this can include substances or machines, methods of work and other aspects of work organization).

Risk: Risk expresses the likelihood that the harm from a particular hazard is realized.

Thus it can be seen that the statutory definitions of hazard and risk are much narrower than those of the CITB and the legal duties are therefore **minimum standards of performance**.

The CITB's definitions would be much more suited to the TQM/loss control/continuous improvement approach described in earlier chapters.

Hazard	Risk	Score			Control
		H	M	L	

Hazard: Something with the potential to cause harm.   H = High; M = Medium; L = Low.
Risk: The likelihood that harm is realized.

Figure 6.1 Basic risk assessment requirement

Instructively, Latham, writing in 1994, some nine months before the CDM Regulations came into force, saw them first as a factor likely to influence quality and second as a *de facto* step towards the registration of contractors.

Taking the statutory definition as the minimum standard, there are two further components which must be present in risk assessments carried out under Regulation 3 of the Management Regulations, namely:

- they must be suitable and sufficient;
- they must consider the likelihood and severity of the unplanned event actually occurring.

There is no written requirement in CDM Regulation 13 for design risk assessments to be suitable and sufficient, or that likelihood and severity should be considered. However, **Designing for Health and Safety in Construction** makes it clear that it is good practice for designers to include these components in their risk assessment procedures.

Thus the basic requirements of risk assessment are illustrated in Figure 6.1, with a reminder that they should also be suitable and sufficient.

6.2 Design/construction risks: differences examined

The main reason is that, as was shown in Chapter 5, designing, managing and constructing are all different activities carried out by different vested interest groups at different times in the construction process cycle and in different places.

Therefore, the requirements for design risk assessment simply extend the basic health and safety management principle that responsibility rests where it falls, with the person responsible for carrying out a piece of work.

The planning supervisor, a new role created by the CDM Regulations, has statutory responsibility for ensuring that all designers, whoever they may be, co-operate with each other and comply with the requirements of CDM Regulation 13; in other words, that they apply the hierarchy of risk control to their designs, or at least to that part of the overall design of any construction project that they have actually prepared.

In Chapter 5 it was demonstrated that the procurement route and the contract strategy have a major impact on deciding the identity of designers. In Chapter 4, the impact of the project management arrangements and structures was also demonstrated. Against this background, the importance of the CDM Regulations can now be seen fully. They are perhaps the most impressive piece of legislation applying to construction in modern times. Their significance lies in the way in which they are drafted, because they take into account the differences between the six main procurement routes and still afford clients the widest possible range of choices in making the main appointments of planning supervisor, principal contractor and designers, and others in the supply chain who also need to make appointments of designers and contractors. The details of appointment strategies are discussed in Chapters 5 and 7.

Design is the starting point for any process and construction is no exception. The design defines the work to be carried out, and very often, can be crucial in determining how it is to be carried out. Designers will sometimes refute this suggestion, pointing out that contractors are free to decide how they will erect structures once they are awarded a contract, but frequently the design must be assembled in a particular order and sequence that effectively limits contractors to a few, or even a single, methodology. The designer is best placed to understand and manage this.

It is often the case that designers can do more to eliminate hazard and risk at the drawing board, than the contractor on site. Sometimes, designers are the only people who can do anything about hazards.

Designing for Health and Safety in Construction is published by the Construction Industry Advisory Committee of the Health and Safety Commission and makes it clear that a systematic approach to design risk assessment is required of and by designers.

Recent reports R172 and R173 published by CIRIA (1998) also make it clear that a rigorous design risk assessment procedure is an essential part of CDM compliance.

Unfortunately, most of the published works on CDM avoid describing systems – they usually deal with approaches – of design risk assessment.

There is a reason for this. Designers must decide for themselves what systems they put in place for risk assessing their own design, and because they are competent, they are the people best placed to decide what these systems should be.

If they were to rely on somebody else's system, for example, one contained in a CIRIA guide or a CITB or HSE publication, the system may in certain circumstances be less than is required to deal competently with the hazards and risks in a particular design, or piece of design. It also explains why planning supervisors must not tell designers what to do, either in terms of designing or design risk assessment.

Nevertheless, if a designer had followed a published system, he or she may be able to put this forward as a defence in any proceedings, pleading that the system failed, not them. In turn, this might lead to some form of proceeding against the writer or publisher of the system, and collectively, these reasons go most of the way towards explaining why nobody has so far put forward the definitive design risk assessment procedure. Also, HSE have always made it clear that they do not encourage a 'tick list' approach towards risk assessment.

The writer is not about the break with precedent but feels, based on experience of emerging best practice, that he can go some way towards offering better guidance on design risk assessment methodology than has been made available to date. However, the strongest possible warning is given that the following is not a system to be followed religiously on all jobs, and to the extent that any reader follows it, it is given in good faith, but may not be adequate for the circumstances of a particular project or aspect of design.

6.3 Design risk assessment

At Appendix 4, pages 40 and 41 of the CDM Regulations and Approved Code of Practice, guidance is offered on the format of **The Health and Safety Plan prepared under Regulation 15 (1–3)**. The Appendix shows that the Health and Safety Plan is set out under nine separate headings, namely:

- Nature of the project
- The existing environment
- Existing drawings
- The design
- Construction materials

- Site-wide elements
- Overlap with client's undertaking
- Site rules
- Continuing liaison.

In discharging his duties, it is very useful if the planning supervisor can persuade designers to present design risk assessment information under the nine separate headings, because this allows information to cascade from designers straight into the health and safety plan.

Of the many questions to be asked of designers by planning supervisors, one obvious question will always be '*What unusual hazards or risks does your design contain under the headings of, the existing environment, or the design, . . . and so on.*' Reference to the subtext accompanying each heading then gives designers further insight into how they might be expected to address the detail of design risk assessment.

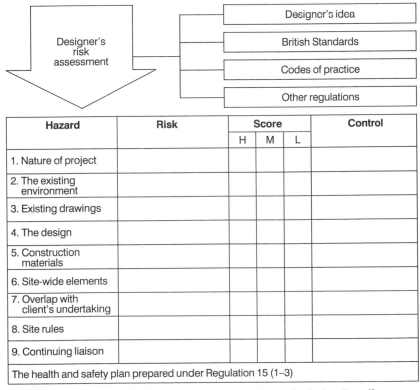

Hazard	Risk	Score			Control
		H	M	L	
1. Nature of project					
2. The existing environment					
3. Existing drawings					
4. The design					
5. Construction materials					
6. Site-wide elements					
7. Overlap with client's undertaking					
8. Site rules					
9. Continuing liaison					
The health and safety plan prepared under Regulation 15 (1–3)					

What unusual hazards and risks are present in your design under the heading of?

Figure 6.2 Design risk assessment approach

If all designers can be persuaded to adopt this approach as the minimum standard to be applied, then planning supervisors are able to organize information for inclusion in the health and safety plan in a consistent way. It is also suggested that dutyholders may be able to demonstrate that, at least in approach, they have attempted to comply with the requirements of the approved code of practice.

6.4 What design should be risk assessed?

The definition of design at CDM Regulation 2 is very broad, and includes 'drawing, design detail, specification and bills of quantities.'

Similarly, the definition of designer is very broad, extending to, 'any person who prepares a design ... or arranges for any person under his control to prepare a design.'

In practice, and taking into account the various arrangements shown in Chapter 5, design and designers can include any and all of the following. All of their outputs should be risk assessed all of the time:

- architects and engineers, civil, structural and other;
- mechanical, electrical and public health engineers designing building services;
- interior designers, landscape architects, shop fitters and the like;
- quantity surveyors and other surveyors and specifiers;
- main contractors and specialist contractors either developing consultants' designs for the permanent works, or designing temporary works including formwork, falsework and scaffolding.

Guidance Note 55 of the Approved Code of Practice makes it clear that CDM Regulation 13 stands alone, in effect, that all designs should be risk assessed whether or not it is intended to build the design. Guidance Note 66 appears to contradict this by recognizing that some designs may never be built. Nevertheless, it is clear that the regulators intend all design to be risk assessed to an appropriate standard and level. The important point is that 'design' is not just drawings; it is the family of documents which collectively comprise the information to be used to build the structure.

It is unlikely to include a design brief and if it was intended to include these documents within the definition of design then the words would probably have appeared in the regulation. However, it is quite likely that clients come within the statutory definition of designer, being persons who regularly arrange for persons under their control to carry out design.

6.5 What risks are there in design?

Many people have come to understand, particularly since the implementation of CDM, the full extent of the part played by designers in managing risks on site.

Merely placing a building in a particular location on site can give rise to one complete set of hazards; locating it elsewhere on the same site eliminates the first set, but creates an entirely new set of hazards. Relevant HSC statistics for 1993/4 are given in Tables 6.1 and 6.2.

The question is, what can designers do to improve the situation by design?

Two broad principles apply. First, designers must do what is reasonable for a designer to do when he or she is designing. Second, the duties of Regulation 13 are qualified by what is **reasonably practicable**.

Table 6.1   Construction process with the highest reported injuries

Process	Reported accidents
Finishing trades	5745
Resource movement	2186
Hard surfacing	1573
Ground works	1428
Manual handling	1345
Loading/offloading	1154
Labouring/attendance	868
Bricklaying	789
Structural erection	722
Scaffolding	663

Table 6.2   Construction activities and fatalities: Relationships

Falls from height	56%
Trapped by collapses or overturning	21%
Collision with moving vehicle	10%
Accidents involving electricity	5%
Falling/flying objects	4%
Machining processes	3%
Hot or harmful substances	1%

By way of example, turning to 'Falls from height', Table 6.2 shows that 56 per cent of deaths involve falls from ladders, scaffolds or from or through fragile roofs. Designers can do many things to eliminate, reduce or control the risks to those who will construct, clean, alter or demolish buildings or structures, or falls from height, including:

- providing for the prefabrication of high-level components at ground level;
- use of non-fragile materials in roofing or vertical cladding;
- provision of hand-rails, guard-rails and the like for use during the aforementioned activities;
- provision of access cradles or other equipment;
- making components smaller or lighter for ease of erection, maintenance or replacement.

Many designers might reply, particularly engineers, that these are obvious matters which would normally be addressed during designing. Furthermore, they might go on to say that compliance with British Standards and codes of practice brings automatic compliance, but this is not necessarily the case. In any event, the writer has found in practice that different design disciplines approach design risk assessment in very different ways.

It is probably true to say that engineers do approach safety in a rational and systematic way, because their design discipline demands such an approach anyway. Architects and other designers tend to concern themselves more with concepts and aesthetics, matters which are much more difficult to systemize or analyse, and so their approach to risk assessment tends to be less robust or clear cut. Having said that, some architects do carry out very thorough and competent risk assessments, but these merely highlight how far the rest of the profession has to go to achieve the same standards.

Surveyors, as a group, and quantity surveyors in particular, tend to consider themselves as a non-designing profession, because they are mainly concerned with the preparation of bills of quantities and sometimes specifications. This is an interesting area, which in the writer's view is likely to give rise to a number of test cases in the not too distant future. Some of these issues are examined in Chapter 7. Suffice to say that bills of quantities are firmly included in the statutory definition of design and bills nearly always contain clauses such as the following:

> The contractor shall visit the site and discover everything there is to know about this site, whether it is reasonable to have done so or not, and bear all possible consequences.

Words like this may now be illegal and could result in the writer of such a clause finding himself in prison. Clauses like this are devices intended to pass risk from its creator to someone else, which of course, opposes the basic philosophy of health and safety management, that risk rests where it falls.

6.6 Design risk assessment in practice

Figure 6.3 is a small case study used by the writer at various workshops involving designers. The writer usually shows the sketch on a slide and asks the question 'Are there any unusual design hazards or risks in this sketch?'

Answers given usually indicate that delegates think there are unusual design hazards including:

- collapse of trench sides;
- working in confined spaces;
- materials falling into trench;
- people falling into trench;
- flooding in the trench;
- gas escape;
- possibility of encountering bad ground;
- contamination in the ground.

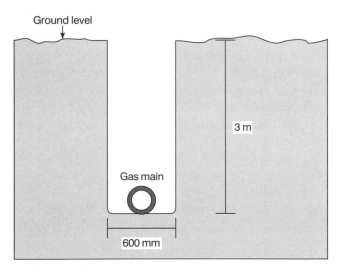

Figure 6.3 Design risk assessment in practice

All of these are distinct possibilities, but the question remains, are these unusual risks created by the designer, and contained in the design?

In the writer's view, to the extent that any of them exist, they are hazards which might arise if the contractor digging the trench and installing the gas main works incompetently. In other words, they are all hazards that could give rise to risk in the event of unsafe work methods on site, but none of them are risks created by the designer or his design. In fact, the sketch illustrates the designer's address of his responsibilities under CDM Regulation 13.

Figure 6.4 shows what the designer would really like to do with the gas main if he could; lay it on the surface. But he knows that he cannot because it will give rise to many hazards if he does, most of which will be high risk. Accordingly, his risk assessment will reveal the following:

Hazard	Risk	Score			Control
		H	M	L	
Collision with moving vehicle	Gas escape	●			Elevate pipe
	Explosion	●			Protect pipe
					Bury pipe
Fracture of pipe due to ground movement	Gas escape			●	Elevate pipe
	Explosion			●	Movement joints
					Bury pipe

Any gas main laid on the surface will obviously be at risk from collisions with moving vehicles, perhaps leading to gas escapes and explosion. The designer can elevate the pipework, but this is likely to be expensive and the collision hazard doesn't go away. It could be surrounded with concrete or a protective cage but this only reduces hazards and probably introduces others in connection with maintenance.

Burying the pipe is the cheapest option and eliminates the hazard. It doesn't introduce any new design hazards and provided the contractor works safely in

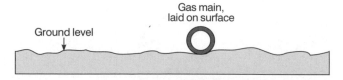

Figure 6.4 Design before application of hierarchy of risk control. From a client's point of view the ideal solution is to lay the pipeline on the ground, because it costs no money to do so

accordance with the regulations that apply, all construction work method risks will be competently dealt with.

But this is only a fraction of the picture. The sketch is merely a small part of what would inevitably be a much larger design and a much greater amount of design detail. Figures 6.5 and 6.6 add further information and another dimension to the matter. The potential for an entirely new and different range of hazards and risks is now introduced into the project, some of which cannot be avoided or designed out. The importance of the guidance offered at Appendix 4 of the ACoP accompanying the CDM Regulations now becomes clearer, along with the format of the Health and Safety Plan prepared under Regulation 15 (1–3). Also, the need to adopt a systematic approach to design risk assessment is made clear.

Systematic approach

At Appendix 6.1, an illustration of the designer's risk assessment is enclosed, based on Figs 6.5 and 6.6. The points to note are as follows:

- Hazards have been identified under the nine separate headings.
- Risks arising from the hazards have been identified.
- Where the designer requires a particular control measure to be taken, or where he has assumed that one will be taken, he has made this clear in his risk assessment.

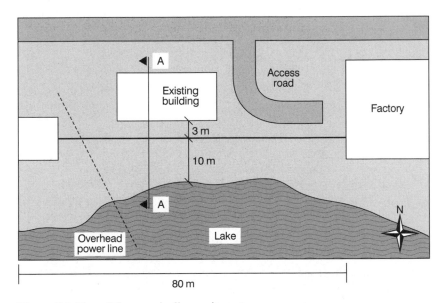

Figure 6.5 Plan of the gas pipeline main route

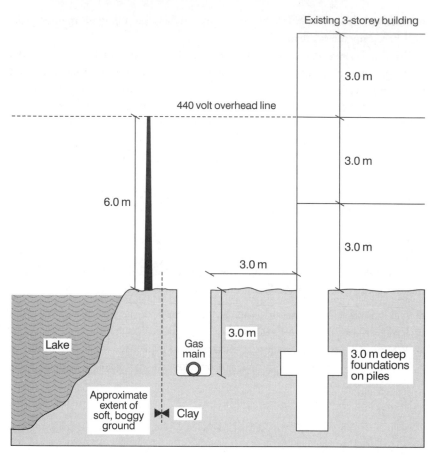

Figure 6.5 Figure 6.6 Section A:A

- He has dealt with the design he has prepared, based on the information he has in his possession at the time the design was prepared. If he changes his design, or if further information becomes available, he might need to review his risk assessment.
- The designer has also assessed the likelihood and severity of the risks his design has created.
- The designer has not identified all risks which might arise. He has merely dealt with those he believes are not immediately apparent to a competent main contractor (or another designer) and has given enough information, in his opinion, for the contractor to make adequate provision and to allocate adequate resources when pricing and executing the works.

The planning supervisor must now take a view, for he must take into account the contents of the risk assessment in compiling the Regulation 15 (1–3) health and safety plan.

6.7 Construction risk assessment in practice

Guidance Note 77 of the CDM Regulations and approved code of practice make it clear that the health and safety plan must be sufficiently developed for it to form part of any tender documentation or similar proposals. This is to enable tendering main contractors, who are also potential principal contractors, to take the content into account when preparing their tenders, or similar proposals, for example where the price is the subject of negotiation rather than a tender competition.

The detailed implications of this are discussed more fully in a later chapter, but remaining focused on risk assessment, there are now two aspects to consider whilst construction work is taking place on site.

- First, the principal contractor must respond to the Regulation 15 (1–3) health and safety plan.
- Second, the principal contractor must develop the Regulation 15 (4) health and safety plan before any work starts on site (see Figure 6.7).

In order to do this, he must carry out two risk assessment procedures. The first one must deal with the design risks set out in the Regulation 15 (1–3) plan,

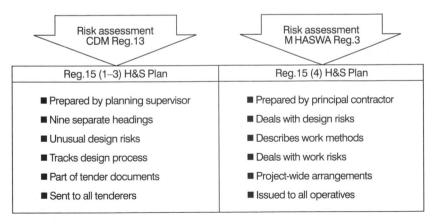

Risk assessment CDM Reg.13	Risk assessment M HASWA Reg.3
Reg.15 (1–3) H&S Plan	**Reg.15 (4) H&S Plan**
■ Prepared by planning supervisor	■ Prepared by principal contractor
■ Nine separate headings	■ Deals with design risks
■ Unusual design risks	■ Describes work methods
■ Tracks design process	■ Deals with work risks
■ Part of tender documents	■ Project-wide arrangements
■ Sent to all tenderers	■ Issued to all operatives

Figure 6.7 Regulation 15 (1–3) and Regulation 15 (4) health and safety plan: one document in two parts

and the second one deals with the risks arising from the contractor's own chosen methods in carrying out the work on site.

The purpose of this is to ensure that there is a continuous risk assessment regime applied to all CDM compliant construction work throughout the lifecycle of the project.

6.8 Construction risk assessment: legal requirements

The legal requirements for risk assessing all workplace activity, including work on building, civil engineering or process engineering sites, are set out in the Management Regulations 1992 and ACoP, at Regulation 3: Risk Assessment.

The essential requirements of a risk assessment are:

- It must be suitable and sufficient.
- It must deal with any person involved or affected by the work activity however they are employed.
- If more than five persons are involved in a work activity, any risk assessment must be in writing.

Risk assessment responsibilities fall on employers and upon the self-employed.

Suitable and sufficient

A risk assessment is suitable and sufficient when it achieves the following:

- It identifies the significant hazards arising out of the work.
- It identifies risks arising from hazard.
- It ascertains the likelihood and severity of risk.
- It sets out a hierarchy of control measures including:
  - eliminate, reduce or, control of identified risk.
- It leads to the design and implementation of a safe work system, which includes a method to monitor, audit and review the arrangements made (see Figure 6.8).

Thus a methodology for dealing with risk assessment is set out in Figure 6.9. It will be noted that in all important respects, it is similar to Figure 6.1. In other words, the approach to assessing risk in construction is the same as in design.

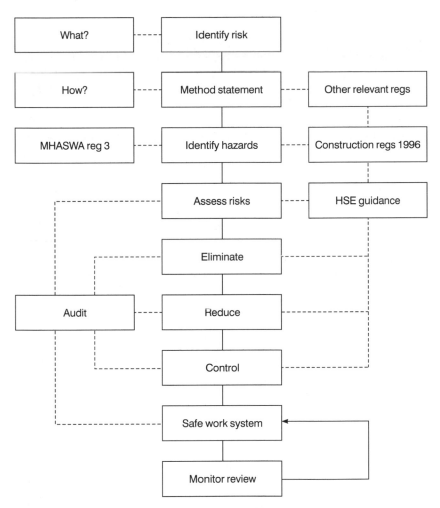

Figure 6.8 Construction work method risk assessment

6.9 What risks are there in construction?

Again, the contents of Tables 6.1 and 6.2 are directly relevant and these arise out of the chosen work methods of main contractors, subcontractors and the self-employed. There are many risks, most of which are foreseeable and preventable, because the majority of tasks at the work face tend to be repetitive and relatively straight forward, for example, bricklaying, plastering and carpentry and joinery. Some of these activities require higher skill levels than others, but there is nothing inherently unsafe in any work activity involving a properly trained and competent operative.

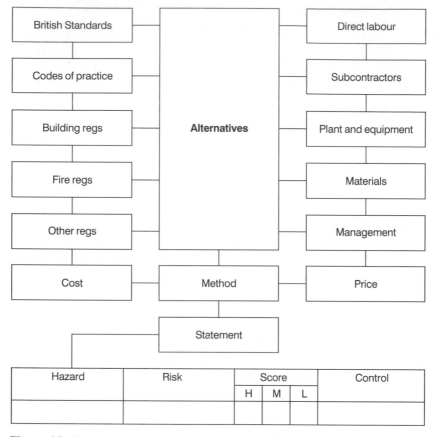

Figure 6.9 Construction risk assessment methodology: five or more persons

Consideration of the statistics in Tables 6.1 and 6.2 reveals the following. Finishing trades have more accidents in terms of number but this is because there are generally more people involved in finishing works than any other activity. Falls from height account for 56 per cent of fatalities.

Put the two together, and most experienced observers might agree that the sort of accident waiting to happen might involve a plasterer working off an extemporized arrangement of scaffold planks over a stairwell, or a carpenter working on a steeply pitched roof with no safety harness and no hand-rail edge protection.

Finishing trades are generally taken to include:

■ plastering and screeding
■ floor and wall tiles

- floor coverings
- painting and decorating
- glazing
- suspended ceilings
- joinery second fix
- electrical second fix
- mechanical second fix
- plumbing and sanitary second fix

A safe work system will prevent these sort of accidents from happening, but not on its own. It must be interwoven into the contractor's working methods and can be considered alongside the quality argument. Any work method that allows operatives to work in the manner described above is likely to lead to a large number of non-conformances or defects in workmanship or materials performance, as well as accidents. In many cases, accidents happen because operatives have failed to follow instructions, but this is an excuse rather than a reason. It is still management failure.

Behind the bare statistics, there are several things that drive the construction process that go to the root of the poor safety record.

- **Competitive tendering**: Most construction projects are put out to competitive tender and despite claims to the contrary by clients, for the most part the only thing that matters to them is price – the lowest price. Tenderers know this and so the only bidding strategy that works is to submit the lowest price. Thus, the competitive tendering process is a procedure to procure a price, and the price represents the contractor's best guess at the amount of money he needs to quote to win the contract. In many cases, it will not even be the main contractor's price, in the sense that he will not have calculated it himself. Most of the effective work on the tender will have been done by subcontractors and suppliers.

  There is no correlation between this price and the cost plus a reasonable mark-up of actually carrying out the work. Thus, for the most part, contractors' subcontractors and the self-employed – and lately consultants – are pricing jobs in a manner that requires them to work unsafely because they cannot afford to allocate adequate resources to undertake work properly in the way they know they should. If they do, they know their bid will be unsuccessful.

- **Subcontracting**: Statistics (HSE) show that 51 per cent of the workforce is self-employed, but most experienced observers will say that at the workface the figure is much higher. Many national contractors employ hardly any direct labour. Everything is subcontracted and the same applies in building, civil engineering, process engineering and house building. The competitive tendering/risk passing/subcontracting regime has fostered a blaming, buck-passing culture that relies on the cynical exploitation of a workforce which is declining rapidly in terms of numbers, motivation and skill levels. This has arisen because margins in the industry, at around 1 per cent of turnover, are not high enough to permit adequate education and training programmes, and for the past 10 to 15 years the workforce of the industry has largely been left to fend for itself.

- **Risk passing**: Another name for competitive tendering is adversarial leverage, and when this purchasing strategy is used, it leads to adversarial relationships. These are further characterized by onerous contracts which pass risk from clients along the supply chain. Frequently, the risk ends up with those least able to manage it, and because it is badly managed, or not at all, it happens. It is as simple as that.

A disgruntled client once famously said of construction that it is 'a cottage industry that deals in millions'. It is the writer's belief that he probably had these matters in mind when he made that remark.

6.10 Construction risk assessment: case study

It might be instructive to illustrate risk assessments of construction by way of a worked example. Upon receipt of the tender enquiry, tendering main contractors will have received a copy of the Regulation 15 (1–3) health and safety plan, and will study it, along with the drawings, specification and the rest of the tender documents.

What happens next is a matter for each contractor. There is no right or wrong way of either pricing or risk assessing the job, and it is very difficult if not impossible to deal with the two issues separately.

Consider the excavation of the trench and the laying of the pipeline (see Figures 6.5 and 6.6).

Methodology

During the estimating and planning stage, most contractors will produce a method statement along the following lines:

- clear site of undergrowth and shrubs;
- set out line of trench;
- strip top soil and set to one side;
- excavate trench in, say, 10-metre long sections;

- support sides and de-water;
- install pipeline and make joints;
- test installed sections with air;
- backfill trench sections;
- reinstate top soil;
- connect to factory meter;
- connect to live mains in compressor house;
- purge main;
- commission and test;
- clear site.

This is the contractor's equivalent of a design, and, just as in design, there are a number of options. For example, the contractor has decided to open up 10 metres of trench (out of 80 metres) at any one time, lay 10 metres of pipe, backfill and reinstate. He could have chosen any length, but experience tells him 10 metres is the amount he can expect to excavate, lay and backfill in a day and this method eliminates, reduces or controls a number of hazards and risks, including trench collapse, exposure to adverse weather, vandalism, people falling into the trench and so on. To an experienced competent professional, this decision would come quickly and easily in the particular circumstances of this project. This in itself is a risk assessment, but is it suitable and sufficient? No, it is not.

Two further issues are important; first, the choice between hand and machine excavation, and, second, the choice between directly employed labour or subcontractors.

The two are inextricably inter-linked – with each other, with the method statement and with health and safety management.

There is about $240\,m^3$ (500 tonnes) of material to be excavated and backfilled. The application of standard estimating tables gives the following approximations:

- Hand excavation and backfill would take about 1000 man hours at a cost of about £7000.
- Excavation and backfill by a JCB-type machine and driver would take about 9 days at a cost of about £150 per day, total £1350. The JCB could also be used for other things, such as distributing pipes, when not digging.
- Excavation by a larger tracked backacter machine would be quicker and might save a day on site, but would be more expensive and the machine would be less versatile. Eight days on site, including driver, might cost £175 per day, total £1400.

Hand dig would, therefore, be eliminated on time and cost grounds before any serious consideration was given to health and safety. The choice would narrow to JCB or tracked backacter, and would probably fall to the former, not just on cost grounds but on the better all-round flexibility of use of the JCB-type machine.

Contractors rarely write these decisions down, and certainly not in risk assessment form, mainly, it is believed, because they are seen as business rather than health and safety issues.

Similarly, the use of subcontracting as opposed to directly employed labour is often seen as a simple commercial choice. The subcontractor would be expected to provide his own machine, insurances and fuel and to bear all of the risks of plant failure, adverse weather, unforeseen ground conditions and the like. If bad ground were encountered, would it really be the responsibility of the digger driver to take measures to prevent the sides collapsing on to the men in the trench?

People who believe it is the digger driver's responsibility appear regularly before the courts pleading their case, only to find that the law and judges do not always agree with them. Unfortunately, by the time they discover this, somebody has usually been killed or injured by a trench collapse. Another aspect of the subcontracting issue is information and training of the digger driver. Much would depend on **how** the digger driver and his machine are employed, as was examined in Chapter 5.

Having dealt with these issues, the next step involves a detailed address of risk assessment. Some contractors might include the foregoing in their risk assessment; it is axiomatic that the majority probably would not.

Most contractors will start off by risk assessing their own method statements, taking into account the contents of the health and safety plan. Specific aspects of it might be dealt with as a separate issue, but for the most part, this would not be the case. Accordingly, a contractor's risk assessment might take the form shown at Appendix 6.2. He would move systematically through his method statement taking into account relevant regulations and HSE guidance. The main set of regulations that will always be relevant are **The Construction (Health, Safety and Welfare) Regulations 1996** but in this case study, others include:

- Construction (Lifting Operations) Regulations 1961;
- Highly Flammable Liquids and Liquefied Petroleum Gases Regulations 1972.

Figure 6.9 illustrates the construction risk assessment process, with particular regard to the way in which other regulations feature in the process.

The cardinal point is before any risk assessment can be carried out, it is necessary to make clear statements about **what** is to be done (the WBS in a project management structure) and **who** is to do it (the OBS in a project management structure). Cost will always be an issue. Although there is no statutory requirement to consider costs in any risk assessment, it is expected that it will be.

6.11 Method statements

There is no statutory requirement for contractors, or anyone else, to write method statements. It is, however, fairly common to find that there are contractual obligations placed upon contractors and subcontractors, and sometimes designers, to do so.

The use of method statements is encouraged by HSE and occupational safety and health practitioners for several reasons.

1 In getting people to write things down, it encourages them to think about the task in hand.
2 It encourages them to commit to what they are writing.
3 It helps communicate the planner's thoughts and intentions to operatives.
4 It serves as a basis for co-ordination with other activities, and for planning.
5 It establishes an audit trail.

Most people recognize that method statement writing is a skill to be encouraged for all of the right reasons; they are very useful 'tools' for communicating information between planners and doers.

The question of detail, how much or how little of it, to include in a statement always arises. Again, there is no right or wrong answer to the question, although the CDM Regulations, in particular Regulation 9: Provision for Health and Safety, require contractors (and designers) to be able to show that they have allocated, or will allocate, adequate resources to managing health and safety in their work activity. In practice, this cannot be considered separately from the production of method statements.

So, the question then arises, is there a minimum performance standard which must be achieved in writing method statements, and demonstrating adequate resource? Indeed, is there any link between the two and how is this relevant to risk assessment.

Risk assessment must be 'suitable and sufficient', for that is the minimum standard set by the law. Hopefully, it has been demonstrated in this chapter that in order to carry out a suitable and sufficient risk assessment, a suitable and sufficient method statement is an essential prerequisite. If there is no method statement, then there is nothing to risk assess.

There is no set format for method statements, but the one shown above is typical of those used in construction and has proved adequate to compile a suitable and sufficient risk assessment of the way in which the contractor proposes to work. So there is a clear link between method statements and risk assessments, and the example quoted here would probably meet the requirements of Regulation 3 of the Management Regulations.

But it still leaves a problem in terms of demonstrating adequate resources, and compliance with CDM Regulation 9. CDM ACoP Guidance Note 36 makes it clear that adequate resources includes 'the necessary plant, machinery, technical facilities, trained personnel and time . . .'.

So should method statements make reference to plant, materials, technical facilities, etc.? Most dutyholders would reject the idea on the grounds that it creates unnecessary paperwork and the resultant documents would probably not be read by anyone anyway, or would be superseded quickly if things changed on site. These are valid arguments.

The counter argument is this. The gas pipeline job will be priced on the basis of assumptions and calculations made by the estimating team at the time of tendering or pricing the job. For example, they will have considered the issue of hand dig at a cost of £7000 and 1000 hours (which could be anywhere between 25 working weeks for one man or two and a half working weeks for ten men if they all work 40 hour weeks), a JCB at 9 days and £1350, or the track machine at a cost of £1450 and 8 days.

These are all matters that require the use of resources in the form of plant, machinery, technical facilities, trained personnel and time, and the estimating team will have noted and documented their decision somewhere.

These must now be passed on to the team who will carry out the work on site, and this will almost certainly not be the estimators. Consequently, it is good business practice to communicate the pricing criteria to the site management team.

It is astonishing how infrequently this seemingly simple, common sense step is actually taken in construction. It is commonplace for the tender price build up to be withheld from the site manager on the grounds that this information is 'commercially sensitive' or that he 'doesn't need to know it'. Often the real reason is that the estimators have guessed the price or used subcontractor prices without examining them, or put in a deliberately low bid in an attempt to secure the contract, and they do not want to expose themselves to criticism from the team that must now live with the consequences of the estimator's decisions and actions in underpricing the job. Everybody knows that underpricing leads to under resourcing, which in turn leads to under performance and unsafe, unprofitable working. There is a plethora of information available to contractors on how jobs should be priced, and the CIOB, which operates the Chartered Building Company Scheme, publishes its Code of Estimating Practice (sixth edition) which is the standard work on building contracts. Consequently, there is no reason why *bona fide* contractors should not comply with the code, as a minimum standard.

Given that common sense and good business practice prevails, then the method statements shown earlier might now be amended to show the following:

> When compiling a method statement, any writer would be well advised to have alongside him or her a copy of the Construction (Health, Safety and Welfare) Regulations 1996 for it is essential that all work on site is carried out in compliance with them. In this case study at least 23 out of 35 regulations are directly relevant and three schedules out of ten apply.

Expanded method statement

- **Clear the site of undergrowth and shrubs** using JCB driver and banksman and two 15-tonne wagons, all material carted to tip off site; 2 days.
- **Set out** line of trench; engineer and chain man plus instruments; 1 day.
- **Strip top soil**, and set to one side. Assume 80-m long × 3-m wide strip, 150 mm deep. Allow for JCB driver and banksman 2 days, spoil to be stacked in heaps on site and kept separate from trench arisings for future re-use.
- **Excavate trench** in 10-m long sections. Starting at factory end and working back towards the compressor house, allow eight sections and all excavated material is to be stacked alongside but 2 m away from the trench as work proceeds. Allow for JCB, driver and banksman, supervised by engineer; 1 day per section.
- **Support sides and de-water the trench** as work proceeds. Allow for three 3-m long Heras hydraulic trench box sections to be lifted into place as work proceeds by the JCB with the specialist timber man in attendance at all times. Box sections to be jointed from the inside and aligned by engineer using an instrument before allowing pipelayers into the trench. Maintain a 50 mm diaphragm pump on site at all times under the control of the timberman. The pump shall be located alongside, but not in, the trench.
- **Install pipeline and make joints.** Two pipelayers will work in the trench and the JCB and banksmen will be at their disposal using chains and slings to lower pipes into the trench. The pipelayers will bottom up the trench as work proceeds and lay an 80 mm sand levelling bed prior to placing the

pipe. When ready, the pipelayers and the engineer will align the pipe before it is jointed to the previous pipe section using a MIG uPVC welding machine.

- **Test the installed sections** with air. This will be done by the two pipelayers using test procedure 42A. No test should be put in hand until 30 minutes after the last joint has been made and an air test of 200 psi must be held for 15 minutes. The trench shall be evacuated during the test.
- **Backfill trench sections.** Upon successful completion of an air test, the free end of the pipeline shall be protected. The trench box at the free end will be left in place and the others shall be removed by the timberman and the JCB and banksman. The trench shall be evacuated during this procedure. When removed, trench boxes will be laid on their sides along the north side of the trench, and shall be checked and maintained by the timberman. When the sections are removed, the JCB attended by the banksman and the pipelayers shall backfill the trench using the excavated material stockpiled alongside. Material will be placed in the trench in layers of about 300 mm. It must not be dragged and dropped. The digger will compact each layer as work proceeds and the pipelayers will compact the completed backfilled trench with six passes of a hand operated 5-tonne vibrating roller.
- **Trench excavation.** Pipelaying and backfilling is intended to be completed at the rate of one 10 m section per gang day.
- **Re-instate top soil.** Upon completion of the trench and the successful purging of the pipeline, the JCB, driver and banksman shall re-instate the stockpiled top soil, spreading and levelling it along the line of the backfill trench.
- **Connect to factory meter.** The specially trained two-man jointing gang will attend site when all straight pipelengths are laid and backfilled. Two open pits will be left, fully supported at each end of the line, approximate size 2 m × 2 m × 3 m deep. At the factory end, the jointing gang will bring the gas pipe vertically up the wall and out of the trench, and 900 mm above ground level will form a hole through the factory wall, take the pipe through the hole and connect it to a new gas meter inside the factory. When the connection is made, the supports will be withdrawn and the pit backfilled.
- **Connect to live mains in compressor house.** The jointing gang will then move to the other end of the pipeline and form a connection in the manner described above. However, a permit to work system will be implemented, operated by the site manager and Transco, and the final connection to the live main will be made by Transco's operatives.
- **Purge main**. This should be done by the jointing team under a permit to work procedure set out in Transco's standard procedures.

- **Commission and test.** Upon successful purging and certification by Transco, the mains should be commissioned by the jointing team in accordance with Transco's standard operating procedures.
- **Clear site.** Whilst purging, commissioning and testing is in progress, the engineer, JCB and banksman shall clear the site of all debris, remove the Heras fencing and site cabins and leave it clean and tidy.
  (*Note*: The above is for illustration purposes only.)

So the essence of good method statement writing is as follows:

- It should state what is to be done.
- It should describe how it is to be done.
- It should say what time has been allowed.
- It should indicate in broad terms the labour, plant and materials to be used.
- It should describe any work sequences or relationship to preceding or succeeding activities.
- It should take into account the contractor's risk assessment.

6.12 Who carries out risk assessments?

The guiding principle is, the person who controls the way the work is to be done at the workface, should undertake risk assessment. In construction, this is usually the site manager, but he will rarely be in sole charge of the workplace, particularly on a large or complex project.

Similarly, in the writer's view, the guiding principle on method statements is that they should be prepared by the person who controls the way the work is to be done at the workface, and the same caveat applies.

The purpose of health and safety plans is to bring clarity to this situation, because the principal contractor is given statutory responsibility for ensuring that all work is risk assessed and that method statements are prepared, and these comprise the basis for developing the plan.

Within the parameters of goal setting regulation, the parties are free to decide for themselves who physically undertakes risk assessment, and the preparation of method statements, and one supposes that this is an essential freedom which must be granted in a free market economy. However, in the writer's view, within a TQM or QA environment, it is impossible to divorce the pricing of jobs from the management of them.

If this view is accepted, then estimators and others responsible for tendering or pricing construction work must ensure that a risk assessment/method statement regime is included as an integral part of tendering procedures.

This should be made clear to site management teams, who should then be given clear procedures for developing risk assessment/method statement documents received from estimators. These should build on the work done at tender stage and take into account changing circumstances as they occur on site, and all new information received as work proceeds. The resulting information should be included in the Regulation 15 (4) health and safety plan.

The situation to be avoided at all costs is one where the site manager is presented with a bundle of drawings and documents a week before work is due to start on site, and is given no support, no briefing and little or no insight into the basic assumptions upon which the tender price has been calculated, but is told to get on with it. It happens regularly and frequently in construction.

6.13 Case study conclusion

Clearly, this method statement is somewhat longer than its predecessor, but not excessively so, and it contains far more information; enough, hopefully, to enable the site manager to understand the way in which the work has been priced whilst leaving scope to manage the job in the way he wishes within the time, cost and quality standards set out in the contract.

It is still not a complete statement of the works and would normally be read in conjunction with, for example, a Gantt chart type programme. However, it probably does demonstrate that adequate resources have been, or will be, allocated to dealing with the health and safety issues identified in the health and safety plan and contractor's risk assessments.

Avid readers will want to check for themselves, less avid readers might wish to accept the writer's assurance, that the new method statement takes into account:

■ the issues raised in the design risk assessments and health and safety plan at Appendix 6.1
■ the control measures recommended in the contractor's risk assessments at Appendix 6.2
■ the Construction Regulations 1996.

Consequently, all of the elements of a safe work system are now in place, and the point is made that there is a clear correlation between risk assessment, method statements and adequate resource. Furthermore, there is a clear correlation between these and the contractor's business management arrangements, particularly his estimating systems, and the way in which essential tendering or pricing information is communicated from the estimating team to the site management team.

Contractors who rely on extensive use of subcontractors, and use lump sum prices in their tender price build ups, will for the most part, find it impossible to demonstrate how they have allocated adequate resources to managing health and safety on site, because the underlying reality is that they have not allocated any resources, they are relying on their subcontractors and the risk passing mechanisms in the contract matrix to allocate the necessary resources.

Now there is a clear audit trail back to clients who employ contractors of this type.

6.14 Other regulations requiring risk assessment

On 1 January 1998 there were a total of nine sets of regulations which required risk assessment. Two of them have been examined in detail here, namely the Management Regulations and CDM. The other seven are:

- Manual Handling Operations Regulations 1992
- Personal Protective Equipment at Work Regulations 1992
- Health and Safety (Display Screen Equipment) Regulations 1992
- Noise at Work Regulations 1989
- Control of Substances Hazardous to Health Regulations 1994
- Control of Asbestos at Work Regulations 1987
- Control of Lead at Work Regulations 1980

There are other regulations requiring risk assessment but they are highly specialized and are not commonly occurring, for example, Ionizing (Radiation) Regulations.

HSE have published two easy-to-read guidance notes:

- Five Steps to Risk Assessment;
- A Guide to Risk Assessment Requirements.

Between them, these two documents deal with all of the common and essential features of risk assessment in the broad workplace – that is, not specifically construction – but are of great practical use to anyone wishing to look further into the subject.

6.15 Chapter summary and conclusion

The risk assessment philosophy has been in place since 1974, but it was not given practical effect until 1992 when the Management Regulations were introduced.

The essential elements of risk assessment are that they must be suitable and sufficient, deal with likelihood and severity, and embrace all those involved in or affected by the work activity being assessed. If more than five persons are involved, they must be in writing.

There are five steps to be taken in carrying out risk assessments and they are:

1 identify hazard;
2 assess risk;
3 apply the hierarchy of risk control;
4 monitor and review;
5 audit.

The statutory definitions of hazard and risk are minimum standards which all employers and the self-employed must achieve.

Before a risk assessment can be carried out, it is necessary to consider carefully and in some detail the method of work to be adopted, and any alternatives.

Although there is no formal requirement to do so, it is good practice to produce written method statements, dealing with how, what and how long, as well as with labour plant and materials.

In construction, there are two aspects to risk assessment, namely:

■ CDM, which requires designers to risk assess the design of construction and the management arrangements to be made;
■ the Management Regulations, which require contractors to risk assess the way in which they intend to carry out construction on site.

The common features of both are the same, namely hazard, risk, control, high, medium and low, but the type of hazards created by designers is very different from those created by contractors.

Very often designers can do far more to control safety on building sites at the drawing board than contractors can achieve on site, where they may be constrained by the decisions designers have taken.

The nature of design risks, and how they are generated, remains poorly understood and is an area where further research is urgently needed. Many designers do not meet the minimum standards required of them by law.

The standard of method statement writing and construction risk assessment among contractors is low and is heavily influenced by:

■ competitive tendering;
■ subcontracting;
■ risk passing.

A competently written method statement, based on a thorough design risk/ construction risk assessment is an excellent management tool and an effective means of communication.

At present, there are nine commonly occurring sets of regulations with a risk assessment requirement, and the two most important are the CDM Regulations and the Management Regulations.

Appendix 6.1 Project: new gas pipeline: design risk assessment

Hazard	Risk	Score			Control
		H	M	L	
Overlap with client's undertaking					
	The gas pipe is being installed by Transco to service the factory of ACME Products across land owned by the Environment Agency. A licence has been obtained from the EA.				The contractor must liaise with: ■ Transco ■ ACME Products Ltd ■ Environment Agency Details of their specific requirements are set out at Section 1.0 of the specification.
	Access to the compressor house is by special arrangement with Transco.				Details of Transco's access arrangements to the compressor house are given at Section 2.0 of the specification.
	Access to the factory is by special arrangement with ACME Products. ACME make high-quality toiletry products where biological contamination can lead to major problems with product quality and food hygiene conditions apply to the area where the gas main enters the building.				Details of ACME Products, access arrangements and work environment arrangements are given at Section 3.0 of the specification.

Site rules

Site rules			
Work in the compressor house and the factory of ACME Products shall be carried out under a permit to work system.			Contractor to design and implement a permit to work system for: ■ compressor house ■ ACME Factory
Method statements are to be submitted to the contract administrator prior to starting work on site for: ■ Excavation of trench ■ Protection to the foundations of the existing building ■ Installation of the pipeline ■ Commissioning and testing of the pipeline			
The lake is to be used for an angling competition in the first two weeks of June, Contract Weeks 8 & 9, and 500 competitors and 10000 spectators are expected. Unauthorized access to the site could be a problem.	■ Injury and damage to persons or property including operatives and members of public.	●	An adequate hoarding or protective fence to be continuous between the compressor house and the factory.

Hazard	Risk	Score			Control
		H	M	L	
Construction materials					
	No construction materials have been specified with any known health hazard.				
Site-wide elements					
	The existing access road is assumed to be the main access on to site.				We will try to arrange a licence to construct a temporary access between the existing building and the compressor house. Otherwise, contractors will use the existing access road, and shall design a traffic management system taking into account the needs of other users.
	Contractors' temporary site area shall be adjacent to the access road and shall be approximately 30 m × 30 m.				The exact location will be allocated when the result of enquiries into the temporary access road are known.
	Loading and off-loading areas shall be included in the above.				The contractor will be required to fence off the entire enclosure and provide lockable gates.

Hazard	Consequence		Control measures
Traffic and pedestrian routes.			Must be dealt with in accordance with the 1996 Construction Regulations.
Existing environment			
Undermining of adjacent building.	Subsidence, or foundation failure resulting in damage to property or injury to people's health.	●	The existing building is on piles, so the risk is low. No control needed.
Electric shock from overhead line.	Electrocution from contact between digger and OHL affecting driver.	●	Contractor to use gates and banksmen.
Vehicle collisions on access road and junction with main road.	Injury to persons, plant and property including operatives, visitors and members of public.	●	Consider temporary access road between existing building and compressor house. Time restriction on use of access road for contractors' plan and deliveries.
Exposure to existing hazardous materials.	Ill health of operatives.	●	Asbestos survey in compressor house and factory.
Flooding of trench from watercourses or 'perched' water.	■ Ill health of operatives. ■ Injury leading to drowning of operatives.	●	■ Probe line of trench with steel rod. ■ Dig trial pits. ■ Check LA records.

Hazard	Risk	Score			Control
		H	M	L	
The design					
Contamination of the pipeline during installation.	■ Explosion.		●		Purge the pipeline before commissioning.
	■ Malfunction of equipment or process in factory affecting operatives.		●		
Pressure fluctuations in gas supply.	■ Explosion.		●		Design commissioning and testing procedure.
	■ Malfunction affecting members of the public.	●			
	■ Explosion.	●			Use an 'outage'.
	■ Interruption to supplies affecting other workers.		●		
A no smoking policy shall be implemented on site for the duration of the project. No fires shall be lit on site.					
All plant used on site shall be spark proofed.					
No personal stereos or radios shall be permitted on site.					
All contractor vehicle parking shall be within the designated compound.					

All heavy vehicles leaving site shall pass through a wheel wash, provided by the contractor.				
The access road and the main road shall be kept clear of all contractors' plant materials and equipment and all debris and mud shall be cleaned immediately.				
Continuing liaison				
The main contractor shall design any necessary temporary works. Details of these shall be submitted to the contract administrator six weeks before any work is put in hand on site. A design risk assessment shall be submitted with the design. A copy of this shall be sent to the planning supervisor.				
If unforeseen ground conditions are encountered, the main contractor shall immediately notify the contract administrator and the planning supervisor. If the conditions are such as to warrant a change in the design, work shall cease until a new design is prepared and risk assessed. This will then be issued to the main contractor who will review methods, making any necessary changes and carrying out new risk assessments as and where necessary.				

Appendix 6.2 Project: new gas pipeline: contractor's risk assessment

Hazard	Risk	Score			Control
		H	M	L	
(i) Trench excavation					
Collapse of sides.	Injury to operatives.		●		Shore up sides. HSE SS60.
Buried cables or gas pipes.	Electrocution of operatives.		●		Safe digging practice. HSE SS7.
Work in confined spaces.	Poor ventilation at workface.		●		No engines in trench. No solvents in trench. No welding in trench.
Use of digger as mobile crane.	Overturning. Overloading, affecting driver and workforce operatives.		●		Instruction for Operator HSE Sheet 19.
Access/egress to trench.	Falling, slipping ladders.	●			Correct ladder use. HSE Sheet 2.
Excavated material stockpiled alongside trench.	Material falling into trench.	●			Keep material at least 1 m away from trench side, along south side of trench.

HSE, SS68 etc. These are HSE guidance issued through HSE and re-issued at regular intervals. They are considered 'best practice'.

(ii) Trench excavation

Hazard	Consequence			Control measure
Pipe sections falling into trench	Injury to operatives in trench	●		Lay pipes on cradles 2.0 m away from trench
Pipe jointing leading to gas build up	Poor ventilation affecting operatives	●		Use polypropylene welding/jointing machine
Gas escape in compressor house when jointing new pipe to supply	Explosion injuring operatives and members of the public		●	Implement a permit to work system.
Damage to pipe during backfilling	Fracture/leakage leading to lowered pressure and malfunction or explosion affecting factory workers or members of public.		●	Back fill in layers not exceeding 300 mm and compact as work proceeds. Air pressure test on each section each day.

7

Construction (Design and Management) Regulations 1994: practical application

7.1 Introduction

In the case of *Regina* v *Associated Octel Co Ltd*: HL: 14 November 1996, the following circumstances emerged.

> For some years, the appellants had used a small firm of contractors to undertake certain repairs at a large chemical plant. Eight of the contractor's employees were engaged more or less full time on the site and operated a 'permit to work' system. This required them to fill in a form for every job, stating what was to be done and obtaining authorization from the appellants' engineers. The engineers then decided what safety precautions were required and issued a safety certificate which imposed the conditions under which the work was to be carried out.
>
> This system was part of the statement of health and safety procedures which the appellants were required to draw up and submit to the Health and Safety Executive. Whilst undertaking repairs to a tank, one of the employees was badly burned in a flash fire and the appellants were convicted of a breach of the Health and Safety at Work Act 1974 section 3(1), which imposed a duty upon the employer to 'conduct his undertaking' in such a way as to ensure, as far as reasonably practicable,

that the persons in his employment who might be affected would not be exposed to a risk to their health and safety. Octel were fined £25 000. They appealed and the Court of Appeal dismissed it so they appealed to the House of Lords. The appellants argued that the incident had not been caused by the way in which they conducted their undertaking within the meaning of that section of the Act, as the contractors were independent contractors and they had no right to control the way in which they did their work.

In dismissing the appeal, the court held the following:

'. . . a person conducting his own undertaking is free to decide how he will do so. Section 3 requires the employer to do so in a way which, subject to reasonable practicability, does not create risks to people's health and safety. If, therefore, the employer engages an independent contractor to do work which forms part of the conduct of the employer's undertaking, he must stipulate for whatever conditions are needed to avoid those risks and are reasonably practicable. He cannot, having omitted to do so, say he was not in a position to exercise any control . . .

The concept of control as one of the tests for vicarious liability, serves an altogether different purpose. An employer is free to engage other employees or independent contractors. If he engages employees, he will be vicariously liable for torts (i) committed in the course of their employment. If he engages independent contractors, he will not. The law takes the contractual relationship as given, and, in some cases, the control test helps to decide the category to which it belongs. But for the purposes of Section 3, the category is not decisive.

The question, as it seems to me, is simply whether the activity in question can be described as part of the employer's undertaking . . . it is part of the undertaking not merely to clean the factory, but also to have the factory cleaned by contractors. The employer must take reasonably practical steps to avoid risk to the contractors' servants which arose, not merely from the physical state of the premises . . . but also from the inadequacy of the arrangements which the employer makes with the contractors for how they do the works . . .

As the question of whether having the tank repaired was part of the conduct of Octel's undertaking was . . . one of fact, it should properly have been left to the jury. Even if, as I think, the only rational answer was 'yes', it should still have been left to the jury . . . The tank was part of Octel's plant. The work formed part of a maintenance programme planned by Octel. The men who did the work, although employed by an independent contractor, were almost permanently integrated into Octel's larger operations. They worked under the permit to work system. Octel provided their safety equipment and lighting. None of these facts was disputed. In these circumstances, a properly instructed jury would undoubtedly have convicted.

This case illustrates a number of points about health and safety law, and the CDM Regulations in particular, although the writer hastens to add the prosecution was not brought under CDM.

First, the defendant, a large company in the petrochemical sector, apparently felt so strongly about the particular circumstances of the case that, having lost it in the first instance, they took it to the Court of Appeal and then to the House of Lords; procedures which would have cost them far more than the £25 000 that they were fined.

On the face of it, Octel had entered into an agreement that is replicated across the entire spectrum of UK business and industry all of the time. They had 'out-sourced' a task, an essential part of their business process, to an apparently competent, adequately resourced, specialist contractor. They had managed the contractor's activities within their own overall process. Octel appear to have held the view that they had certain responsibilities towards the contractor, but it seems that they did not believe that they could be held liable if things went wrong. In the writer's opinion, if that was Octel's view it would be shared by many others, but not on the evidence before them, by the courts.

In the opinion of three courts, Octel was in control of the workplace. In effect, it was telling the contractor and his employees what to do by way of the permit to work system, and the fact they were almost completely integrated into Octel's process.

The second point to be made is this. The out-sourcing arrangement made by Octel is mirrored on every construction project throughout the UK but for the most part, with far lower standards of health and safety management than those applied by Octel as a matter of routine. The importance of the CDM Regulations in construction is that they place every client on every CDM-compliant project in the UK in potentially the same position as Octel, but in most cases, without that organization's in-house health and safety management expertise, and with contracting organizations that in many cases they do not know and have not worked with before.

Furthermore, many of these companies are in effect one-man bands. Remember, only about 12 000 out of 200 000 contractors in the UK construction industry employ more than seven people. It is still possible to own and operate a contracting company on £2 paid-up capital.

7.2 Dutyholders and impact of the regulations

The CDM Regulations came into force on 31 March 1995 and immediately created the following duties:

- clients;
- designers;
- planning supervisor;
- principal contractor;
- contractor.

Two statutory documents are created:

- the health and safety plan;
- the health and safety file.

The regulations apply to **construction work** on **projects which contain structures** and generally cover all but the smallest of jobs.

Applicability

The definition of construction work is given at CDM Regulation 2, and has been made as broad as possible in an effort to close any loop holes. In broad terms, it includes:

- building, new build and maintenance;
- civil engineering, new build and maintenance;
- process engineering, new build and maintenance;
- all demolition;
- site investigation;
- enabling works;
- services installations;
- cleaning.

The regulations do not talk about contracts, they talk about projects. There can be one contract or many in a CDM-compliant project, and Figure 7.1 describes some of the circumstances that can occur.

Figure 7.1 illustrates a typical speculative development project, for example a block of shops in a town centre. Normally, the developer would borrow most of the money on a short-term arrangement from a bank, hoping to sell the completed, tenanted properties as soon as possible to an investment institution

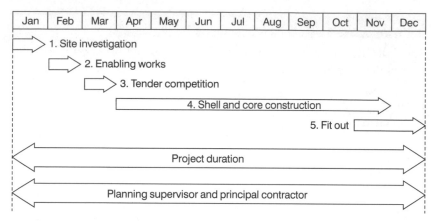

Figure 7.1 CDM and projects

such as a pension fund. In good times, capital gains to be made on developments such as this can be huge, but timing and time are always of the essence and developers will always seek to squeeze time and drive capital costs down to the bare minimum.

For these reasons, they will often use some sort of 'fast track' construction. Design and build is very popular with developers (see Chapter 6).

In this case study, the developer might typically appoint an architect and a consulting engineer in January to prepare the **employer's requirements**. They would need a site investigation to discover the nature of the ground, location of existing services, possible contamination and so on. A contract for that might be let straight away, because the consultants would need the results before they could proceed.

An enabling works package might then be undertaken in February. This might comprise demolishing any redundant buildings on site, diverting services, removing contaminated ground, site clearance, and the like. The purpose of this is to eliminate surprises and reduce risk in the much bigger shell and core contract to follow. In March, the employer's requirements might be issued as part of a tender competition to, say, six design and build contractors, and in April the successful contractor would be expected to start work on site, finishing in November.

As work proceeds, the developer will be searching for tenants, and if he finds them, he will agree specific fit-out packages with them for mechanical and electrical services, shop fronts, floor coverings and the like. Either the tenants might undertake this work themselves, using their own contractors, or the developer might do it for them and rentalize the capital cost of the works to make it easier for the tenant to move in. Either way, these works might be undertaken as a series of separate contracts, perhaps one for all of the shop units, or as a series of separate contracts for each unit.

In the CDM sense, this is one project comprising at least four contracts. The project starts when the site investigation starts and it ends when the last tenant fit-out is finished. The **client's** obligation is to ensure that a **principal contractor** and a **planning supervisor** are appointed for the duration of the **project**, and the start time of the project begins when **construction work** starts on site and ends when construction work finishes.

Figure 7.2 describes a different, but still commonplace, situation. Here, the client might be an NHS trust or a higher education or further education body such as a university. Generally, such clients are experienced and procure construction all of the time. The chart describes what might be a major refurbishment project in an existing building, perhaps several hospital wards or a research laboratory. Typically, the client would break up the work into a series of packages and would manage all of the packages itself, the default design and manage route (see Chapter 6). The chart depicts eleven separate contracts in one project which starts in January and ends in December.

Some people have argued that the opposite is the case; that there are in fact eleven mini-projects, but this is demonstrably wrong. For one thing, it would require eleven separate principal contractors, planning supervisors, health and safety plans and health and safety files and would make nonsense of the duties

	Jan	Feb	Mar	Apr	May	Jun	Jul	Aug	Sep	Oct	Nov	Dec
Demolition	■											
Electrical		■	■				■	■				
Heating		■	■				■	■				
Plumbing/sanitary		■	■				■					
New partitions					■	■						
Suspended ceilings							■					■
Medical gases							■					
Floor coverings										■	■	
Tiling							■	■				
Painting/decorating									■	■	■	
Soft furnishings												■

Figure 7.2 CDM and projects

of co-operation. For another, each contract is part of the same project, and not some standalone matter.

Having established that there is a project which contains construction work, the next step is to ask the questions: 'Is the project notifiable and do the regulations apply to it?'

The notifiable test is given at Regulation 2. A project is notifiable if its construction phase will be:

■ longer than 30 days, or
■ involve more than 500 person days of work.

A day means one normal eight-hour shift and persons engaged on construction work includes professional, managerial and technical staff working on the project, not necessarily from the same organization, but visiting site in connection with the work in hand.

The applicability test is given at Regulation 3. If, at any one time, less than five people work on site, the regulations do not apply. In other words, if five or more people are on site then the regulations do apply.

In practice, clients and their professional advisers must apply both tests at the very early stages – inception and feasibility – of a project. When it is ascertained that:

■ there is a project which contains construction work,
■ the project is notifiable,
■ the regulations apply to the project,

then everybody in the assembled supply chain which will deliver the project has the duties prescribed to their particular role by the regulations.

Turning back to Figure 7.2, and the case study, an experienced person will immediately surmise that, taking the eleven contracts as one project, the project is longer than 30 days and will probably involve more than 500 person days of work, so it is notifiable. Furthermore, there will be more than five persons on site, so the regulations apply.

Therefore a planning supervisor and a principal contractor must be appointed for the duration of the project, and all dutyholders must comply with their obligations under the regulations. There are criminal penalties that can be applied to all of the dutyholders for breach of the regulations.

Some of the prosecutions and penalties from March 1995 to August 1995 are set out below. In that time, nobody had gone to prison for breach of CDM, but some very heavy fines were made and people acquired criminal records.

Sample CDM prosecutions

- An architect's practice was fined £500 for failing to inform the clients of their duties under the CDM Regulations.
- A client was fined £2500 for allowing construction work to proceed without an adequate health and safety plan.
- A client was fined £2000 for failing to provide information about the state or condition of premises at which construction work was to be carried out.
- A designer was fined £3000 for failing to ensure that his design included adequate information about health and safety.
- A principal contractor was fined £2000 for failing to produce an adequate health and safety plan.

The significance of these cases is, first, that they were brought under the CDM Regulations and, second, the cases did not involve notifiable accidents. Some other recent cases brought under general health and safety law which did involve notifiable accidents are outlined below:

- BP was fined £750,000 for failure of safety precautions at Grangemouth.
- In the 'Port of Ramsgate case', fines against four defendants totalled £1.7 m.
- Nobels Explosives were fined £100 000 and £250 000 respectively.

7.3 Client's duties

Clients are given duties at the following regulations:

- Regulation 4, Clients and Agents of Clients;
- Regulation 6, Appointment of Planning Supervisor and Principal Contractor;
- Regulation 8, Competence of Planning Supervisor, Designers and Contractors;
- Regulation 9, Provision for Health and Safety;
- Regulation 10, Start of the Construction Phase;
- Regulation 11, Client to ensure information is available;
- Regulation 12, Client to ensure health and safety file is available for inspection.

The overall aim of the regulations is to actively involve clients in the management of health and safety in their construction projects. However, it is recognized that not all clients are competent to play an active role because they have little or no experience of construction, so they are given an alternative.

Regulation 4

Clients must be competent to undertake the role, but if they are not, they can appoint a client's agent, that is a person or firm who can 'stand in their shoes' for the duration of the project, subject to the proviso that the client's agent must be competent and adequately resourced.

The client must, therefore, ask for a demonstration of competence and resource before making any appointment, and should remember that there is a clear audit trail which can be examined against the regulations and approved code of practice.

Regulation 6

Client or client's agent must make two statutory appointments at the appropriate times. They are:

- planning supervisor;
- principal contractor.

The client/client's agent must ensure that both appointees are competent and adequately resourced to undertake the roles, and the appointments must be made in sufficient time to enable the planning supervisor and the principal contractor to discharge their duties properly. In most circumstances this will involve:

- appointing the planning supervisor at inception, or very shortly after;
- appointing the principal contractor before any construction work starts on site.

The earlier both appointments are made, the more benefit and value for money there is to be found.

Regulations 8 and 9

To the extent that the client/client's agent appoints any designers, they must also be competent and adequately resourced and the person making the appointment must check that this is the case. The extent to which clients appoint designers is heavily influenced by the chosen procurement route, but as a rule of thumb, any consultant employed by the client for the purpose of preparing tender or similar documentation will be a designer, if the client arranges for him to prepare drawings, design details, specifications or bills of quantities. This can often include contractors, through clauses such as the following:

> The contractor shall submit proposals for the design, construction and operation of a structure that will meet the client's service provision requirements . . .

Regulation 2 has this to say on the matter:

> In determining whether any person arranges for a person . . . to prepare a design . . . regard shall be had to the following, namely –
>
> A person does arrange for the relevant person to do a thing where:
>
> 1 he specifies in or in connection with any arrangement with a third person that the relevant person shall do that thing . . .

Regulation 10

Clients must not allow construction work to start on site unless a health and safety plan complying with Regulation 15 (4) is in place. If they do so allow, they are breaking the law and can be fined and jailed (see sample prosecutions). This is a highly visible offence. Either there is a compliant plan on the day the Health and Safety Executive asks for it, or there isn't. If there is not, it is not possible to quickly cobble together some document that will get the inspector off your back. Plans are developed over time and require input from a number of individuals and organizations. Their development can be traced along the audit trail, so it is much easier and safer to ensure that a compliant plan is produced.

Regulation 11

Clients must ensure that information is provided, or made available, about the site or premises where construction is to be carried out. This is an interesting regulation all round, but particularly so from a contractor's point of view. It has already been mentioned in this book that clients often write clauses into contract along the following lines:

> The contractor shall visit site and carry out his own site investigation. He shall discover everything there is to know about this site whether or not this is reasonable, and shall bear all possible consequences . . .

The purpose of such clauses is to pass all risk in unforeseen ground or premises conditions to the contractor, and insofar as any contractor accepts such an onerous obligation, the clause is effective. It does just what is intended in contract.

However, the position in statute might now be found by the courts to be somewhat different, because CDM Regulation 11 works in the following way. The planning supervisor has to take a view. Has the client provided enough information about the site or premises where the project is to take place to enable designers to prepare a safe design?

There are several points of note:

■ the regulation places duties only on clients and planning supervisors;
■ the regulation requires the planning supervisor to ascertain the information needs of designers. Designers can include consultants and/or contractors.

The type of information that might be required is very broad but could easily embrace:

■ geotechnical, contamination and gas surveys;
■ asbestos and other hazardous materials;
■ structural stability;
■ dimensional and spatial details;
■ trees, water courses and other natural features;
■ existing and previous use.

Designers need to know of these, and other matters, in order to produce a competent design. They also need to know of them to address their obligations fully under CDM Regulation 13. Consequently, planning supervisors must continue to deal with Regulation 11 rigorously as the design develops, and they must press designers to fully identify their information needs, and clients, to provide the necessary information on an on-going, as-needed basis.

To the extent that clients rely on clauses similar to the above, and do nothing about Regulation 11, there is now a clear audit trail back to both the client and the planning supervisor. Many people believe that there may be some interesting court cases in future years involving this regulation.

Regulation 12

Once a project is complete, but before it is handed over, the client must be given the health and safety file by the planning supervisor. Throughout the lifecycle of the building or structure, the client must first make the file available to anybody who might need it, and second keep the file up to date.

Those who might need it can include:

- cleaners;
- maintenance personnel;
- consultants, designing alterations;
- contractors, building alterations;
- demolishers, knocking down parts of the building;
- the client's employees.

Over the lifecycle of buildings, a great deal of construction work can be carried out in the way of:
- planned and cyclical maintenance;
- repairs and replacements;
- alteration or extension.

Every time a piece of CDM-compliant work is carried out, the client has a statutory duty to ensure that the file is brought up to date.

Who is 'the client'?

Finally, the identity of the client is cardinal to the discharge of the duties. The client is defined as 'any person for whom a project is carried out . . .'

The use of the word 'person' is deliberate, for health and safety law envisages that within every organization there is someone who represents its heart and mind and that person is accountable for the organization's address of its legal duties. In practical terms, the role of the client is simply an extension of the role of employer, established under HASWA, but with specific application to construction. One of the main aims of the Health and Safety Executive in creating the role was to bring clarity to areas of possible uncertainty. In a building contract, the name of the employer might well be given as 'the Metropolitan District Council of Gladstonebury', and that is fine as far as the contract goes because it has a corporate identity which renders it visible to its trading partners.

But the Metropolitan District of Gladstonebury is a place, and its district council is a disparate group of elected members and as such, it has no 'heart and mind' because it is not a person. The people behind the title are in fact the Mayor, Burghers and Members of the Metropolitan District of Gladstonebury, but they might not be considered competent in the health and safety sense, consequently it is commonplace for one of the paid executive officers of the Council to stand in the shoes of the Mayor, Burghers etc. and to be its heart and mind.

There are many similar examples including national, regional and local government, universities, multi-funded development projects, and this further explains the role of the client's agent. Where the identity of the client, the person for whom the project is carried out, is not clear, then a client's agent should be appointed.

7.4 Designer's duties

Designer's duties are given at Regulation 13 which, in conjunction with Regulation 16 (see later in this chapter), comprise the central 'spine' of the regulations.

Regulation 13 requires designers to:

- inform clients of CDM;
- apply the hierarchy of risk control to their designs;
- co-operate with other designers;
- co-operate with the planning supervisor;
- provide information about their design for inclusion in the health and safety file.

Inform clients

This envisages that clients will usually (but not always) turn to consultant designers when thinking about procuring construction, and is a sort of fail-safe device to ensure that clients are told about their obligations under the regulations. However, no construction project can proceed without design, and at least one designer, so clients cannot plead that they did not know CDM because, whoever the lead designer might be, he or she would have a duty to inform the client. This includes consultants or, in a design and build situation, the main contractor.

Apply the hierarchy of risk control

The hierarchy of risk control in design is:

- identify hazard;
- identify and assess risks arising from hazard;
- eliminate, reduce or control the risk.

So, the duty is not to design safely or eliminate all risk from design, but to eliminate, reduce or control risk to an extent that is reasonably practicable in the design, or those parts of it actually prepared by the designer. A suitable design risk assessment methodology was considered in Chapter 6.

Co-operate with other designers

The definition of a designer given at CDM Regulation 2 is very broad and, again, uses the word 'person'. A designer is 'any person who prepares a design or arranges for any person under his control to prepare a design'. So it is possible to be a designer without actually putting pen to paper, or sitting at a drawing board.

Designers are likely to be consultants, main contractors and specialist subcontractors, but can also include most other types of subcontractor, suppliers and operatives including the self-employed and also clients. It is commonplace for consultants to prepare concept and outline designs which are then added to by contractors, detailed by subcontractors and modified by the acts of operatives.

Clients can often change their minds about designs whilst they are being built. Consequently, it makes sense for designers to co-operate, but because of the adversarial nature of certain procurement routes (see Chapter 5) it is equally commonplace for them not to do so in practice. Again, the regulations recognize this and in an approach that is admirably and demonstrably free of concessions to vested interest or vanity, treat all designers, whoever they may be, exactly the same in requiring them to co-operate with each other to the extent necessary to discharge their obligations, that is, to apply the hierarchy of risk control to the developing design.

The number of designers and designing organisations is generally a function of the procurement route (see Chapter 5), and the way in which they co-operate and communicate with each other flows out of the type of project management structure used on the job (see Chapter 4).

Figure 7.3 illustrates the designing process in construction. The points to be noted are:

- Different groups of designers are brought on board at different times in the process.
- The appointments and activities of the different designer groups can be separated both in time and by location, sometimes by years and thousands of miles. Getting them all together at the same time and in the same place, if needed, can be very difficult.
- Broadly speaking, each designing group, first, must deal with the work contained in its scope of services and, second, must address the needs of the various external interest groups influencing the project. It must also take into account the needs and wishes of the other groups of designers.

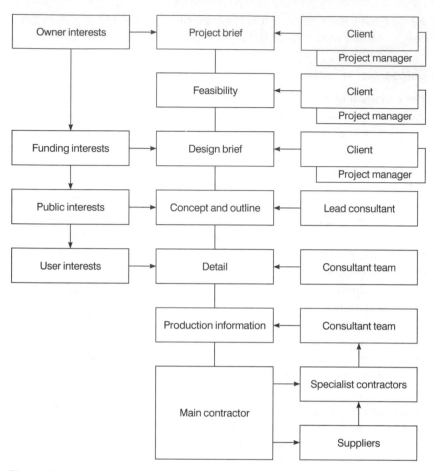

Figure 7.3 Design process: traditional route

Figure 7.4 illustrates the point by way of example, listing the main designer locations and timescales for a new combined heat and power plant in Belfast, Northern Ireland built between 1993 and 1997. To the best of the writer's knowledge, the full complement of designers was never brought together in one place at one time, and many of the appointed designers had no idea of each other's existence.

Design started by one designing organization in 1993 was actually completed by another organization in a different country in 1997, having had significant contributions made to it by two other completely different designing organizations in 1995 and 1996.

The original design was heavily modified, once in 1994 to take into account planning and environmental issues and again in 1996 due to changed client

Project		New CHP plant	
Location		Belfast	
Value		About £5 million	
Timescale		1993 to 1997	
Client		Airframe manufacturer	
Contract		Service contract bespoke	
Contractor		London based	
Architect	Belfast	1993	1994
Structural engineer	Glasgow	1994	1997
Services engineer	Nottingham	1995	1996
Planning supervisor	Cambridge	1995	1997
Quantity surveyor	None appointed	–	–
Main contractor	London	1996	1997
Civils contractor	Dromore	1996	1997
Structural steel contractor	Drogheda	–	1997
Mechanical services	Belfast	–	1997
Electrical services	Dublin	–	1997
Power source	Helsinki	1996	1997

With the exception of the planning supervisor, all of the above carried significant design responsibilities.

In addition, about 30 subcontractors were employed on the job in various capacities ranging from painting and decorating, roofing and cladding and fixing only of reinforcement. Some of these were *de facto* designers.

Figure 7.4 Time and space chart

needs. Parts of the design were modified by contractors undertaking the work in question, sometimes through the formal mechanisms of the contract, and sometimes seemingly because they felt like changing it. The design that was built was very different to the design that was first accepted by the client. Yet the plant worked perfectly when it was commissioned in 1997.

Vested interests between the various designing groups can be a serious barrier to co-operation. A split develops on most projects between consultant designers and contractor designers. The split has its roots in two main, and many lessor, issues. The two main issues are, first, the mindset of some consultants, and second, what is often referred to as 'buildability'.

'**Consultant mindset**' seems to flow out of the way in which the built environment disciplines have been taught in the UK, and is heavily re-enforced by the institutions which represent the vested interests of the various consultant disciplines.

Latham (1994) dealt with some of these issues in a very fair and friendly way in 'Constructing the Team' and flowing from that, CIB Workgroup 9 has produced 'Educating the Professional Team' which points out that the debate on interaction between related disciplines has spanned four decades and is not resolved yet.

Education remains a major issue. Some built environment courses, particularly those dealing with architecture, still have a very narrow focus where students are taught that architecture is an art form and that 'the design' is all important. This seems to encourage some architects to believe that everybody else, including the client, should put the needs of the architect before their own, or the wider needs of the project. Most people employed in construction eventually bump into some bow-tied fellow whose ego outweighs his ability. These people are usually found wanting, not on pure design ability, for many of them are very good at concept and outline, but on the ability to detail their own design in a cost-effective way, provide useable information to enable it to be built, and to administer the building contract. They also seem to suffer from a terminal inability to get on with their fellow men.

The writer recalls one project where he was brought in to troubleshoot a job which was suffering serious time and cost over-runs and major quality problems. It quickly became apparent that the behaviour of the architect, in his dual role of architect and contract administrator, was the prime cause of the project's problems.

At the writer's first meeting with him, he loftily announced he had designed the three most important rooms to be built in England since the Second World War, and that he was surrounded by idiots who didn't understand or share his vision of the concept and didn't have the ability or practical skills to turn it into reality.

On a traditional, fully designed job, this architect was issuing incomplete drawing and design detail to the rest of the team, which he was then expecting them to complete. The team tried hard to co-operate with him, but he treated them with an arrogance that left many of them unwilling and unable to respond. He was in the habit of walking around the job and observing the works. He would wait until a piece of work was finished and would then point out some real or imaginary fault in it and require its removal. He insisted on speaking directly to operatives, usually in a patronising way, but would not speak or listen to their supervisors.

The architect was, quite literally, causing chaos on the job and this meant that the site was unsafe.

On the third day of the appointment the writer realized that this man had to be controlled. When the effects of his behaviour were pointed out to him, he remained icily aloof and simply refused to discuss the issues. When the writer insisted that the architect examine various drawings he had produced to enable the writer to explain to him where and how they were deficient, he refused.

It would have been right to remove him from the job, and to have done so might have solved the co-operation issue at site level but would have created far bigger problems in the legal and contractual environment. He had to be ring-fenced, with an executive architect brought in to stand between him and the rest of the team for the duration of the project.

On a 39 week programme, there was a 30 week over-run. The final account was £3m against a tender sum of £1.95 m; several subcontractors became insolvent and the client avoided arbitration by the skin of his teeth. To this day, the architect refuses to accept that his behaviour, his inability and failure to co-operate with other designers had anything to do with the job's problems.

Happily, such people are not the norm but they do exist and it is very difficult to do anything about them when they are encountered.

Buildability was a major issue in the example quoted above. The entire contracting side of the supply chain was incensed at the impracticable nature of the design information being given to them by the consultants and in particular, the architect. Contractors' specific complaints included:

- poor-quality of information, with essential details including levels and dimensions missing;
- late issue of information;
- changes of mind, resulting in design changes;
- incorrect information, particularly in the mechanical and electrical services, where the architect's drawings might show socket heights 150 mm above floor level, and another consultant's drawings might show them 200 mm above floor level;
- incompatibility of consultants designs with trade practice. For example, a dimension across a window opening and a brick elevation of, say, 1153.5 mm. This does not co-ordinate with standard brick dimensions and it is not possible to cut a brick by hand to within half a millimetre of accuracy;
- lack of co-ordination of information between consultants;
- abuse of the telefax as a means of issuing instruction.

Although the argument was by no means one sided – the main contractor was not blameless – for the most part, the contractors were right. The design team was producing a design that was far too complex, time consuming and costly. The same effects could have been produced far more quickly, cheaply, cost effectively and safely, offering the client better value for money.

The whole project could have been made much more buildable if one designer, the architect, had been prepared to co-operate with other designers, including contractors, to achieve the overall design intent. Unfortunately for whatever reason, the architect was unable to see the contractors in that light, that is, as other designers. It seemed to me, as the person charged with sorting out this mess, that the architect's behaviour stood between him and a clear picture of what was going on on this job.

The design brief called for the finished work to look and feel like an ancient Greek temple. The architect's research and concept and outline design seemed to me to be immaculate. However, the reality was that this ancient Greek temple was being built out of modern materials using modern methods by a main contractor and a team of subcontractors who had won the job in competitive tender on low prices and onerous terms and conditions of contract. The timescale was inadequate and the physical circumstances prevailing within an existing occupied and operational building in the middle of a busy city centre all added to the difficulty and complexity of the job.

At the end of the day, there is only so much to be done with 150×50 softwood, plasterboard, sand cement render and paint to get it to look like a Doric column, particularly if you wish to hide a complicated mechanical and electrical services installation inside the column. The architect was, in my opinion, taking a complex design problem and turning it into a complex construction problem through the production of an over-elaborate design. In practice, this job might have been more successful if the design and build route had been used. There was no lack of interest or commitment on behalf of other designers and contractors involved in the project; they all understood the design intent and were committed to achieving it, but were frustrated at what they saw at the architect's unreasonable behaviour and intransigence. Time and again, I was shown examples of positive practical workable solutions to problems which had been presented to the architect for his approval, only to be rejected on the grounds that they were 'pastiche', 'modernist' or 'utilitarian'.

One example involved the proposed use of sand cement render to achieve a roughcast finish to walls in a public room. The plasterer produced several sample panels to show how this could be done to a standard which seemed to match photographs produced by the architect. However, he rejected them all on the grounds that sand cement was not available to the ancient Greeks, and insisted on the use of a sand lime mix reinforced with horsehair, which he insisted would have been available. However, the plasterers, skilled tradesmen all, had not previously used this material and were unable to reproduce the desired effect. In consequence, this part of the work was delayed for months while the argument rumbled on. One of the plasterers pithily remarked to the architect that he might get what he wanted if he could produce some ancient Greek plasterers. The architect's response was to request his removal from site. To his credit, the main contractor refused to comply.

As previously mentioned, the job appeared to be in chaos with the specialist contractors bearing the brunt of the chaos. The contractor had programmed to work through the job systematically floor by floor, top down, to avoid going back through completed areas, with no more than 12 workfaces open at any one time. By week 30 of a 39 week programme he had 37 workfaces open and had failed to complete a single one. As a result, the site was unsafe and had been rendered so by the difficulties presented to the main contractor and his subcontractors by the incomplete design.

Anyone could walk on the site, including members of the public. On one workface, work had started and stopped six times in six weeks due to incorrect or incomplete information.

On walking around the site, the evidence of unsafe working was there for all to see in the form of incomplete scaffolding, festoon lighting laying on the floor in puddles of water, heaps of debris and builders' rubble at every workface, half-finished work, festoons of cables hanging down below suspended ceiling grids, walkways obstructed by stacks of material, clusters of window frames waiting to go into openings that had not yet been built and so on. All of these are tell-tale signs by which any experienced observer might surmise that control of a job has been lost because the planning and programming systems had broken down.

In this case, they had broken down because the necessary drawings, design detail, specification and bill of quantities was not good enough to sustain them. In other words, designers had failed to co-operate because of consultant mindset, and a design had been produced that was extremely difficult to build. The end result was unsafe working coupled with a completely dissatisfied customer who did not get value for money from the construction process.

It is a sobering thought that similar behaviour to that of the lead consultant might in future be the subject of a criminal investigation on the grounds that the person or organization concerned failed to co-operate with other designers to the extent necessary to demonstrate compliance with the CDM Regulations.

Co-operate with the planning supervisor

Many people have openly questioned the need to make a law in the criminal code requiring designers to operate with each other, and with the planning supervisor, a new role created by the regulations. The reasons are not hard to find.

179

The overall aim of the law is to improve safety in the process of construction, which includes designing and building. The fragmentation of the process has been examined elsewhere, but the point remains of importance here, because designing itself is carried out by two main groups, and within those groups, many subgroups, namely:

- Consultant-designed work;
- Contractor-designed work.

These groups and subgroups share one common factor, namely the contractual environment within which they must perform.

The contractual environment configures not only performance, but also behaviour. The more established forms of building and civil engineering contracts (JCT and ICE respectively) mirror the adversarial system of English law; consequently, they are themselves adversarial forms of contract.

Accordingly, the matrix of contracts used to implement the main procurement routes is, for the most part, exclusive of any requirement to co-operate. The relationships are all linear and singular, that is, between clients and individual consultants, employer and main contractor, main contractor and his individual subcontractors and suppliers. It is instructive to note that in some contracts, the client is only the client of the consultants; he is the employer of the contractor.

So the contract serves its own purpose, that of, *inter alia*, preserving the rights and benefits of vested interest, but it encourages an adversarial approach, when what the process requires is a co-operative one.

Consultant-designed work

The main consultant design disciplines are all represented by a professional institution, RIBA, RICS, ICE, IMechE and so on. The main function of each institution is to further the interests of its members and the profession in general, and the means by which it does so is as follows.

1 Establish the services to be offered by its members.
2 Promote the services to be offered by its members.
3 Publish a standard form of appointment for use by clients and members, and recommended fee scales.
4 Establish and publicize good practice guidance for use by clients and members.
5 Identify and promote continuing educational and professional development requirements for members.
6 Represent members' interests at national, regional and local level.
7 Arrange and promote professional indemnity insurance.

This last point is crucial. All of the institutions publish standard forms of appointment describing the services to be offered by members and recommended fee scales. Generally, the services fall into two broad categories (see for example, RIBA CE95) namely, standard services and additional services, all of which must be paid for by clients at the recommended scale fee, or some other arrangement agreed between the parties. The defined standard or additional services are covered by the member's professional indemnity insurance, the broad terms of which will have been agreed between the institution and insurers.

Any services outside the scope of the standard or additional services will not be covered by the consultant's professional indemnity insurance policy. That is why consultants will not step outside the standard forms of appointment. It is not that they are opposed to giving clients added value for money by including, for example, value management as part of their price, instead, it is to protect their own and the client's interests in the event of negligence in the service delivery. If they get the value management exercise wrong, and are sued by the client, their professional indemnity insurers may not extend cover, and they might either have to meet any claims out of their own resources, or go bankrupt. That is one aspect of the matter.

Another important aspect is that there is no requirement in any of the standard forms for consultants to 'co-operate'. Consequently, co-operation is outside the scope of the agreement and is not covered by professional indemnity insurance. Therefore consultants have no obligation in contract to co-operate and may not be held in breach if they do not co-operate. Having said that, most consultants would immediately say that 'reasonable skill and care' and the requirements of 'good practice' mean that they do co-operate when designing and there is much force in that argument.

However, in practice, what tends to happen is that consultants co-operate to a greater or lesser extent with other consultants. It is questionable whether they co-operate to the same extent and in the same way with designers who are not consultants, that is, contractors.

The example given in the previous section is a clear illustration of this point. The architect seemed to feel that the designing he was doing was somehow different and more important than the designing everybody else was doing. These attitudes probably flow from the people involved and the way in which built environment courses are taught.

There are widely acknowledged deficiencies in teaching standards in the education of construction professionals, and these have been dealt with by Latham (1994) and CIB work group 9 'Educating the Professional Team'.

It is instructive to note that the standard form of appointment for an architect, CE95, has been amended to include co-operation with the planning supervisor as a standard service, in line with the requirements of statute, but there is still no standard of additional service requiring the architect to co-operate with other designers. Thus the role of the planning supervisor in ensuring co-operation between designers now has both a statutory and a contractual aspect.

Contractor-designed work

These, together with co-operation issues, present a different set of problems for designers and planning supervisors in ensuring co-operation. To the best of the writer's knowledge there is no formal educational or training course entitled 'How to be a Contractor'. (The CIOB, which offers excellent and broadly based educational and training programmes to its members, might disagree with the above view, but as a Fellow of the Institute the writer stands by it).

The extent of contractors' involvement in designing the permanent works is a function of the procurement route and the contract strategy. Traditionally, contractors have well-developed skills in the design of temporary works such as formwork systems, falsework schemes (used to support the permanent works) and scaffold systems.

Some of these skills are taught either in colleges of higher or further education, or by the Construction Industry Training Board (CITB), but for the most part, temporary works design is a jealously guarded expertise developed in-house by larger contractors. Sometimes their skills, expertise, and ingenuity achieve world class. The Queen Elizabeth II Bridge at Dartford is an example of advanced temporary works design impacting on permanent works design in a way that enabled the project to be completed nine months ahead of schedule.

The design of formwork, falsework and scaffold is firmly categorized as design (see Guidance Note 71) within the meaning given in the CDM Regulations.

When it comes to designing permanent works, the only group of contractors who can claim any real in-house skills are normally mechanical and electrical services contractors. Most other contractors who design, apart from recognized specialists, appear to rely on a combination of freelancers, consultants, other subcontractors and suppliers. This can give rise to lengthy convoluted and fragmented design supply chains. Ensuring co-operation between these designers is never easy, and frequently it does not occur to the standard required by the CDM Regulations.

It is commonplace to find that a main contractor who has accepted design responsibility for a piece of work simply passes it on to subcontractors. They may either deal with only a part of it, or none at all. This situation does not usually come to light until the operatives arrive on site to actually do the work, whereupon it is discovered that some piece of information is missing. This either delays and disrupts the work, or, the operatives simply do what they think ought to be done, based on past experience.

A regularly recurring scenario involving mechanical and electrical services is as follows. Usually the mechanical services package breaks down neatly into heating, hot and cold water, public health and ventilation. The electrical component breaks down into lighting and power, infrastructure and fixtures and fixings.

However, some sort of electronic controller is nearly always required to regulate the operation of the mechanical installation. It will usually be described in terms of function and performance in the head contract, but the main contractor will frequently omit to make one of the subcontractors responsible for supplying and installing it. The end result is that the heating system does not work when commissioned. It is often the case that this scenario is not seen as a design issue.

When dealing with design, contractors are not usually bound by the same constraints as consultants, with particular regard to standard forms of appointment and professional indemnity insurance issues. Nevertheless they are bound by terms and conditions of contract which can be either a standard form or a bespoke document.

Again it is unusual to find clauses in these contracts requiring contractors or contractor designers to co-operate with other designers, mainly because of the difficulty of identifying a suitable performance standard for co-operation against which to measure 'co-operation'.

In conclusion therefore, the regulators have, by requiring designers to co-operate with each other and with the planning supervisor, introduced an obligation that is not explicitly present within the contractual environment. Most informed observers agree that the requirement should be in the regulations. Against this background, it gives designers and the planning supervisor two major questions to answer.

- What level of co-operation is needed to ensure compliance with the regulations?
- How, when and where is co-operation achieved and demonstrated?

These issues are now addressed in more detail.

7.5 Planning supervisor's duties

The planning supervisor is a new role created by the regulations, and has proven controversial since its introduction in 1995. According to the regulations, and in particular Guidance Note 21, anyone can be a planning supervisor, provided they are competent and adequately resourced.

In practice, it is not quite as simple as that. On all but the smallest and simplest of projects, the planning supervisor is not a matter for individuals, instead it requires a team effort involving a range of skills and expertise which, first, broadly mirror the project design input needed, and second, embrace process and occupational health and safety skills necessary within the specific circumstances of the project.

In Chapter 5, some of the appointment arrangements for planning supervisors were examined, and it was shown that there seems to be a default position for each procurement route, the one that falls naturally into place. However there is a range of fully compliant alternatives available. The type of work, or contract used, can impact on the planning supervisor appointment strategy; Figure 7.4 illustrates the point.

On a traditionally procured project using a JCT contract, the default position is illustrated in Figure 7.5. The lead consultant, usually the architect, is the planning supervisor up to the start of the construction phase, whereupon the main contractor takes over until work on site is finished.

By way of contrast on a traditionally procured civil engineering project using an ICE contract, the default position is illustrated in Figure 7.6. The engineer is the planning supervisor for the duration of the project.

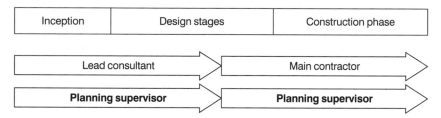

Figure 7.5 Traditionally procured building project

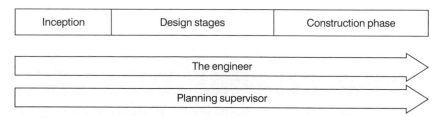

Figure 7.6 Traditionally procured civil engineering project

Thus, the following is relevant:

1 The identity of the planning supervisor can change at any time.
2 The duties of the planning supervisor are spread across the life cycle of projects.
3 The duties remain the same, regardless of the identity of the planning supervisor or the procurement route or the contract strategy.
4 The time at which the duties are delivered can vary.

In broad terms the duties are as follows

■ Ensure HSE are notified of the project.
■ Ensure designers co-operate with each other.
■ Ensure designers comply with CDM Regulation 13.
■ Ensure that any design is risk assessed.
■ Ensure a pre-tender health and safety plan is prepared.
■ Advise clients on competence and resource issues.
■ Ensure a health and safety file is prepared.

Some of the above duties must be done, others apply only if requested.

Notification

Is effected by submission of a Form 10 (Rev) at the appropriate times. Usually, these are first, upon appointment of the planning supervisor, and second, upon appointment of the principal contractor. The form is available free of charge from HSE and is sent to the HSE area office nearest the site.

Designer co-operation

In order to discharge this duty, the planning supervisor must know the identity of all designers. This is a function of the procurement route and the contract strategy, and designers fall into the two broad categories described earlier, namely, consultants and contractors.

The issue for planning supervisors is how to ensure designer co-operation over a period of time, to the extent necessary to ensure compliance.

The starting point, at least in the building sector, is an understanding of co-ordinated project information (CPI). This is based on CAWS – common arrangements of works sections – which is a list of over 300 different types of work encountered in the building process. Generally the preparation of design in the building sector should follow CAWS, with drawings being prepared in accordance with the CI/SfB classification system although it is not mandatory to do so.

The benefit of implementing this system is that the preparation and presentation of design information is, broadly, in line with the way in which works packages are assembled, and with current trade practice in carrying them out, although a note of caution is sounded here; this is not always the case. In general terms, CAWS meets the needs of quantity surveyors and it cannot always be assumed that the needs of trade package contractors are met.

If CAWS is used as the basis of a CPI system, the planning supervisor must choose and implement effective methods which ensure co-operation to the extent necessary to satisfy the requirements of the CDM regulations, and which can be demonstrated to an HSE inspector should circumstances so warrant.

Some methods in everyday use include the following:

- attendance at design team meetings;
- visits to individual designers;
- written design risk assessments;
- correspondence.

The writer's own preferences extend to the first three. A rigorous design risk assessment based on the methodology shown in Chapter 6 (Section 6.3) applied to all design enables relevant information to be collected and exchanged between designers. However, in itself it is rarely enough. Attendance at design team meetings allows the planning supervisor to be part of the team and to understand the developing design to the point where the relevant question can be asked of designers. Visits to individual designers may be appropriate where there is some particularly complex or difficult design issue requiring further investigation by the planning supervisor, or where design responsibilities may be shared a number of different organizations or spread over time.

Figure 7.7 illustrates how the design of a relatively mundane and apparently simple element such as a floor slab, can be shared between five different designing groups.

Element	Component	Architect	Engineer	Main contractor	Subcontractor	Supplier
Floor slab		●				
	Concrete		●			●
	Reinforcement		●			●
	Screed		●			●
	Covering	●				●
	Formwork		●	●	●	●
	Falsework			●	●	●
	Scaffolding			●	●	●

Figure 7.7 Shared design responsibilities (1)

Figure 7.8 illustrates how designing is spread over time, from the identification of the initial need for a floor slab, to its eventual construction.

Correspondence should, in the writer's view, be kept to a minimum. Even if all of these methods are used, it has to be said that there is probably no way in which planning supervisors can ensure co-operation by designers if they do not wish to co-operate. That is probably why the duty on the planning supervisor is to 'take such steps as it is reasonable for a person in his position to take . . . '. This would at least provide a defence to any planning supervisor faced with intransigence by a designer.

In the event of non co-operation, the obvious first step to be taken by a planning supervisor is to draw the matter to the attention of the person who appointed the uncooperative designer. Subsequent steps might involve action by the contract administrator, who may have powers in contract to deal with matters of non-compliance. In practice, the planning supervisor has no such powers and must rely on others to enforce his reasonable requests.

Perhaps the best way of ensuring designer co-operation is for clients to consider avoiding the use of adversarial contract systems, such as JCT and ICE, and the use of co-operative contract systems.

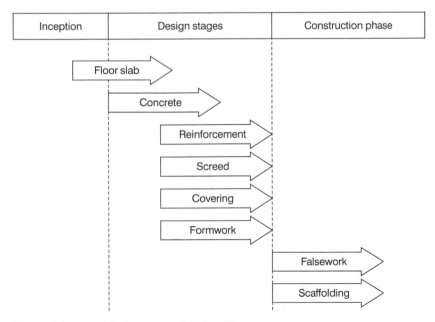

Figure 7.8 Shared design responsibilities (2)

The NEC/ECC form of contract, first published in 1991, is such a system. It has been amended to take into account the 13 requirements of a modern contract identified by Latham (1994) namely:

1 Fairness to all parties.
2 Teamwork based on win/win solutions.
3 Interlocking matrix of documents.
4 Clear language with guidance notes.
5 Separate management and administration roles.
6 Adjudication.
7 A choice of risk allocation.
8 Pre-priced changes.
9 Alternative payment methods.
10 Binding payment dates, interest on late payment.
11 Secure trust funds.
12 Incentives for good performance.
13 Advance mobilization payments.

The GC Works series, whilst remaining firmly an 'employer's' contract also meets the above.

Both ECC and GC Works contracts encourage co-operative behaviour when properly implemented, and assist planning supervisors in discharging their statutory duty of ensuring designer co-operation.

Proponents of JCT/ICE may well dispute the writer's view that they remain beyond the pale of Latham's recommendations. In particular, JCT published amendment 18 (1998) which it says brings all JCT building contracts into line with Latham.

The writer points out that JCT contracts remain firmly rooted in the adversarial system of English law. One clause still in use in JCT contracts has its origins in the 1870s. It takes about 250 unpunctuated words to say what the ECC manages to say in about 30 words.

Designer compliance with CDM Regulation 13

This is another aspect of the co-operation issue. In practice, it is dealt with in the same way as co-operation. Basically, the law assumes that designers will co-operate to the extent necessary to comply with their obligations, and because statute law does not contemplate failure, the planning supervisor is given no remedy if designers fail.

Accordingly, failure is a matter that must be dealt with in contract, for in practice the planning supervisor can do relatively little about it.

Risk assessment of the design

This is the essence of the planning supervisor's role. A methodology was put forward in Chapter 5, and mentioned earlier in this chapter. Guidance Note 55 makes it clear that **CDM Regulation 13, Requirements on Designer**, stands alone. This means that all construction design should be risk assessed, even if there is a possibility, or even a probability that it might not be built, because the reciprocal possibility is that it **might** be built.

Basically, the designer's duty is to show that the design can be built safely. Therefore it is merely good practice to risk assess it, and the nine separate heads of risk assessment set out in Appendix 4 of the approved code of practice; together with the hazards/risks/control methodology, represents the minimum standard of design risk assessment.

The major difficulties faced by planning supervisors include those mentioned earlier, namely:

- Time: the planning supervisor needs to keep track of developing design.
- Fragmentation and location: different aspects of design may be undertaken by different organizations in different places.
- Different standards of design risk assessment.

Some of these difficulties can be dealt with in contract. For example, a standard design risk assessment methodology can be made a requirement of consultants and contractors contracts.

However, it may still take all of the planning supervisor's powers of persuasion to get designers to write their risk assessments out in competent fashion. Hence it is important that planning supervisors are able to form and maintain amicable working relationships with designers, because it is axiomatic that persuasion and encouragement always achieve more than dictat and compulsion.

Preparation of the pre-tender health and safety plan

The planning supervisor must ensure that the pre-tender health and safety plan is prepared; instructively the regulations do not demand that he actually writes the plan. Having said that, it is very difficult for the planning supervisor to convince a client that he offers a value-for-money service if he does not prepare the plan.

The reason for this apparent anomaly is to give clients a broad range of choices within the six different procurement routes and their accompanying contract strategies as to who they actually appoint to prepare the plan.

For example, in a turnkey contract, the client may appoint an independent planning supervisor, but the turnkey contractor will be in possession of all the information to be used in the pre-tender plan, and may therefore be best placed to prepare it, and to submit as part of the tender. Similarly in a PFI/PPP arrangement, the client may use the quality of returned pre-tender health and safety plans as one of the assessment criteria.

However, the practical outcome of this anomaly is that clients must ensure that they place responsibility for preparing their pre-tender health and safety plan with someone, and in most cases, the planning supervisor is the obvious person.

There is no such anomaly within the construction phase plan. It is the responsibility of the principal contractor to prepare it and maintain it throughout the construction phase.

Further details of health and safety plans are given later in this chapter.

Advising of clients on competence and resource issues

These are perhaps the thorniest issues with which planning supervisors must deal. In particular they must understand what is meant by 'competence' and 'adequate resources', and Regulations 8 and 9 and their associated guidance are of key importance.

Only clients can appoint planning supervisors and principal contractors, subject to the provisos that they are competent and adequately resourced.

Anyone can appoint designers and contractors, subject to the same provisos, and in practice a variety of people can and do make such appointments. Sometimes a single organization is appointed to a dual role of designer and contractor. The procurement route configures the process and plays a key part in deciding who makes the appointments, and who is appointed. The regulations recognize this, and set two simple performance standards:

1 The person making the appointments must check the competence and resources of the person or organization that they are appointing.
2 The appointee is entitled to seek advice from the planning supervisor if needed, and the planning supervisor must be in a position to provide advice if and when requested.

It will come as no surprise then that there are no definitions of 'competence' and 'adequate resources', and both will vary from project to project and even from task to task.

However, there is much useful guidance given in the **Management Regulations 1992**, the **CDM Regulations 1994** and various **HSE guidance notes**, including the series of HSE construction sheets numbered 39, 40, 41, 42, 43, 44.

In a broad sense, competence relates to the ability of an organization to comply with their health and safety management obligations within their workplace activity. Workplace activities vary, therefore competence and resources vary.

In the CDM sense, the competence of a designer relates to his ability to comply with CDM Regulation 13, in other words, to co-operate with other designers and with the planning supervisor, to apply the hierarchy of risk control to the design, and to provide information for the health and safety file relevant to the design.

Again the procurement route is important, because it establishes the identity of designers, and configures the supply chain which will eventually deliver the complete design for the project, namely, 'drawing, design detail, specification and bill of quantities'.

Using Figure 7.8 and a concrete floor slab as an example, an architect and a structural engineer will normally be the main designers of such a slab, with the engineer providing the bulk of the detail. Both require different types of competence. The slab may require extensive falsework and scaffolding in order for it to be built safely and this would normally be done by a main contractor or his subcontractors. Again, this requires a different type of competence, but compliance with CDM Regulation 13 is common to all designers.

'Adequate resources' is made clear at Guidance Note 36 of the CDM Regulations. It means adequate 'plant, materials, technical facilities, trained personnel and time'. These must be provided to the extent necessary to implement the dutyholder's identified risk management arrangements. In the case of designers, these flow out of the designer's address of CDM Regulation 13, and in the case of contractors, the contractor's address of Management Regulation 3.

It is a frequent and fully justified complaint of contractors and consultants that clients award contracts on the basis of price only – the lowest price (see Chapter 5). The lowest price mentality of many clients has been examined by Latham (1994) and Egan (1998). The latter has recommended an end to lowest-price-based tender competition. Others, including CIB Work Groups and HM Treasury Central Unit of Purchasing, postulate that price that should never be considered separately from quality, and detailed mechanisms have been put forward by both groups to ensure that adequate regard is paid to issues other than price. Consequently there is more than

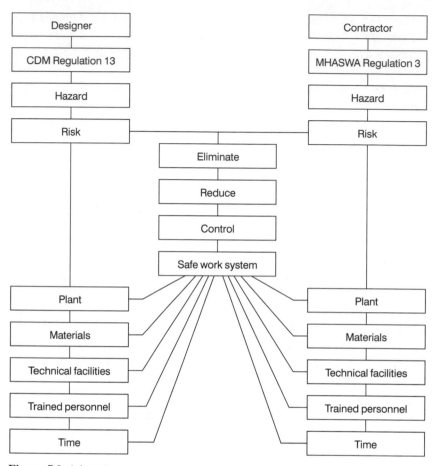

Figure 7.9 Adequate resources

adequate guidance available to clients, and to planning supervisors to enable them to deal with 'adequate resources' on a project and task specific basis.

The competitive tendering process in the UK construction industry in the late 1990s is no longer a competition based upon the cost of doing the job, plus a reasonable mark-up. It is, instead a procedure to procure a price, and the price represents the tenderer's best wild guess of the amount of money that he must quote to persuade the client to award him the contract.

In spending their money in this way clients effectively configure the process. All the supply side can do is react by submitting the lowest price, because it must have work. However, it can do little to add value to its bids, and if it tries, it will almost certainly lose market share to lower bidders.

When attempting to submit the lowest price, bidders (designers and contractors), frequently ignore the relationship between the cost of providing necessary and adequate resources and the price of actually doing the job. The first step is to win the contract, the next is to make the cost of resources fit the price at which the work is obtained, and the third is to seek ways of increasing the price.

The danger in the health and safety sense is that whilst this is going on, designers and contractors may not allocate adequate plant, materials, technical facilities, trained personnel and time to the tasks in hand.

If there is an accident, there is now a clear audit trail through them to the person who appointed them, embracing the planning supervisor to the extent that he advised on adequacy of resources.

Ensuring the health and safety file is prepared

Again the duty on the planning supervisor is not to physically prepare a health and safety file, but to ensure that one is prepared. Essentially the preparation of the file is an information gathering process that can be thought of in three distinct phases.

Phase 1

During the pre-construction stages, the file should contain information of interest and use to designers and may typically comprise site survey and investigation information. This may be held at several locations, or copies of it distributed to all locations. The file itself might comprise an index of all relevant information, prepared and maintained by the planning supervisor, telling designers what information is available, where it can be obtained and how it is accessed. Alternatively, a file could be prepared and held at, say, the lead designer's office. There are a number of ways in which the requirements of the law can be met. The important point is that the client and the planning supervisor must take the lead in deciding and implementing the most appropriate project-specific method.

Prior to starting work on site, the principal contractor must be given a copy of the current file or file index by the planning supervisor, and it makes sense for this to be a formal hand-over meeting.

Phase 2

Once the construction phase starts, responsibility for ensuring the file is prepared remains with the planning supervisor. However in most cases, the principal contractor and his contractors will be the main sources of information for the file, although designers may still be required to contribute on a regular basis. Again the use of an information index by the planning supervisor would satisfy the requirements of the regulations in most cases

Often the developing information can be kept in, say, a fireproof cabinet in the site manager's office as it becomes available.

The file must be complete when the client takes possession of the finished work, and the planning supervisor must ensure that it is handed to the client at this time. This can be difficult, because there is frequently a time lag between completion and hand-over, and the time taken to physically produce essential information such as as-built drawings and operating and maintenance manuals.

For that reason, it is significant that 'time' is included among 'adequate resources' referred to above. Clients must ensure that adequate time is allowed at the end of projects, and before they take possession of the works to enable the file to be completed. Designers and contractors must prepare and present information promptly for inclusion in the file, and the planning supervisor must act promptly to procure the requested information and to maintain the file index. The client must also make a clear contractual arrangement with someone, usually the principal contractor, to physically prepare the file.

Phase 3

Once the client takes possession of the finished work, he is responsible for the file for the lifetime of the building or structure, or at least, for as long as he owns it. The file must be kept up to date, so that every time a piece of CDM-compliant work is done on it, the file is both available and updated. It must also be available to those who use the building or structure.

The planning supervisor's role can be thought of as an active information management function whereby the file is physically prepared by others using an indexing system designed and implemented by the planning supervisor. However there may be circumstances where the planning supervisor prepares the file himself.

Everybody – clients, designers, the principal contractor and contractors – may be required to contribute information to the file, but the planning supervisor must identify the necessary information.

The information may or may not be kept at several locations that may change during the development of the project. If sectional or phased completion is required then the file must be designed and assembled with this in mind, and completion of the file or the relevant parts of it should be a condition precedent to partial, sectional, phased or full completion or possession.

The planning supervisor's responsibility for the file ends when the client takes possession of the building or structure. In order to discharge his duties in respect of the file, the planning supervisor may take the following practical steps:

- At the time of appointment, make clear to the clients the anomalies of the regulations in connection with the file. Ensure that the client places responsibility for the completed file with one organization, usually the main contractor.
- Identify with designers the format and content of the pre-construction file, deciding such issues as to where the file is kept, how it is accessed, what information is to be placed in it and so on.
- Implement procedures to monitor these arrangements, for example, attendance at design or project team meetings.
- Prior to appointing contractors, agree with the client the format and content of the completed file.
- At the time of tender or other arrangement in connection with price, ensure that contractors are clearly told what is expected of them in connection with the file.
- As construction work proceeds, collect information for the file regularly, and review and monitor it and ensure the file is kept up to date. Again the planning supervisor's attendance at site meetings may be a sensible way of doing this, with the site manager being required to keep the file in, say, a fireproof cabinet on site freely accessible to all who might need it.

7.6 The statutory documents

The health and safety plan

The health and safety plan can be thought of as one document in two parts (see Figure 7.10). The first part is the pre-tender health and safety plan, which must comply with regulations 15 (1–3). The important point is that the format of the Regulation 15 (1–3) plan is set out at Appendix 4 of the approved code of practice and forms part of the guidance. It is prudent, therefore, to prepare pre-tender health and safety plans in this format, using the nine separate headings listed at Appendix 4 and taking account of the accompanying guidance.

Clearly if this format is followed, dutyholders can demonstrate attempted compliance with the approved code of practice and the law. Also it is a sensible and highly practical format which enables information to be presented in an organized and user-friendly way.

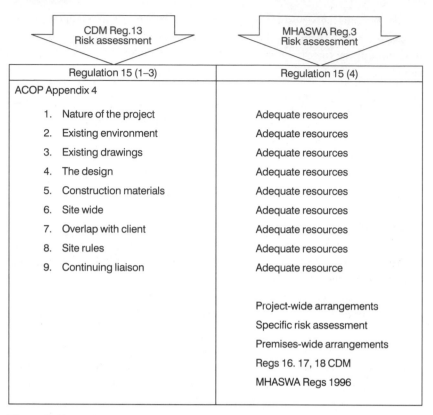

CDM Reg.13 Risk assessment	MHASWA Reg.3 Risk assessment
Regulation 15 (1–3)	**Regulation 15 (4)**
ACOP Appendix 4	
1. Nature of the project	Adequate resources
2. Existing environment	Adequate resources
3. Existing drawings	Adequate resources
4. The design	Adequate resources
5. Construction materials	Adequate resources
6. Site wide	Adequate resources
7. Overlap with client	Adequate resources
8. Site rules	Adequate resources
9. Continuing liaison	Adequate resource
	Project-wide arrangements
	Specific risk assessment
	Premises-wide arrangements
	Regs 16. 17, 18 CDM
	MHASWA Regs 1996

Figure 7.10 Health and safety plan

The pre-tender plan should contain comprehensible information about unusual risks contained in the design, to the extent that they may not be readily apparent in the tender documents, or similar proposals, to a competent main contractor. It is not necessary to describe all risks, merely those which are unusual and which are created by the design. This includes 'drawing, design details, specification and bills of quantities'. Designers can include both those who have prepared design, and those who have arranged for a design to be prepared (see CDM Regulation 2). This can extend to clients, consultants, main contractors, specialist contractors, suppliers and operatives.

Pre-tender plans should be short, as simple as possible and written in plain English. Whilst they should form part of the tender documentation, it is not necessary to write them into the contract as a document. They should be sent to all tendering main contractors, who should ensure that copies are sent to all tendering subcontractors.

At this stage the purpose of the plan is to inform prospective contractors of unusual design hazards, together with the project-wide health and safety arrangements to be made by those who have designed and managed the project to date.

A question that is often asked is 'Are tendering contractors obliged, at this stage, to make any formal response to pre-tender health and safety plan?' The answer is no and yes.

The no answer applies in the sense that is would be unreasonable, and an unnecessary expense, to expect all tendering contractors to respond with a developed health and safety plan and there is nothing in the CDM regulations that requires them to do so.

The yes answer applies in the sense that all tendering contractors are required to note and take account of the contents of the pre-tender plan, and to allow in their prices for allocating the necessary resources to dealing with the unusual design risks, in so far as they impact on their work. CDM Regulations 8 and 9 refer.

When the client intends to place a contract with a contractor or contractors for part or all of the works, he must first ensure that a principal contractor is appointed. In practice the client must then negotiate with the prospective principal contractor to produce a health and safety plan complying with Regulation 15 (4) before he, the client, allows work to start on site. In some cases this might mean either paying the principal contractor to produce the plan or delaying the start of work on site or both.

In order to comply with Regulation 15 (4), the developed plan must firstly deal with the Regulation 15 (1–3) plan. This means that the principal contractor must set out in reasonable detail the arrangements he intends to make and the resources he has allocated to dealing with the unusual design risks, and other issues set out under the nine separate headings of the pre-tender plan.

He must secondly set out in reasonable detail the health and safety management issues that flow out of his own works methods, and he must deal with them in three broad categories:

1 He must set out the project-wide arrangements he intends to make in managing health and safety on site. On all sites this will require an address of the **Construction (Health, Safety and Welfare) Regulations 1996** as a minimum.
2 He must deal with specific risks arising from his, or his subcontractors', proposed methods of work. This will require an address of the **Management of Health and Safety at Work Regulations, Regulation 3, Risk Assessment**, as a minimum.

3 He must deal with any other health and safety issues arising out of, but perhaps not directly connected with, the way in which he intends to do the job. For example, noise generated by work activity can often affect both operatives and members of the public including local residents. This could require an address of many regulations including the **Noise Regulations, Work Equipment Regulations, Asbestos, Electricity** and so on.

The principal contractor must then submit the Regulation 15 (4), the construction phase, health and safety plan to the client, who must then take a view: has the plan been sufficiently developed to enable construction work to start on site (Regulation 10)? If, in the client's opinion, it has been so developed, then work can start, and if it has not, then the client must not allow work to start.

The client may, if he is in doubt, ask the planning supervisor for advice about the adequacy of the developed plan. In advising clients, the planning supervisor may pay attention to Guidance Notes 76 to 88 of the approved code of practice, and particular attention to Guidance Note 84. He must also have knowledge and experience of the contracting process, sufficient to enable him to offer competent advice.

The plan does not need to be complete. It needs to deal in detail with work activity taking place in perhaps the first six to eight weeks of the project. Once a compliant plan is obtained and work starts, the principal contractor is then solely responsible for developing the plan as work proceeds, and before any particular activity starts. However, if significant changes are made by the client or the design team, then the planning supervisor will have continuing duties in connection with the Regulation 15 (1–3) plan, for as long as design continues this part of the plan must also continue to be developed, thus triggering a response in the Regulation 15 (4) plan.

This is why the health and safety plan is one document in two parts. One part deals with design risk, the other deals with construction risks and it should always be kept in mind that health and safety plans are intended to be read by operatives actually doing the works described in them.

Clients frequently complain that it can be difficult to obtain compliant plans from main contractors, particularly in a competitive tendering situation. That is simply one of the pitfalls of the process and attracts little sympathy from contractors. The view is that if you choose adversarial leverage (lowest-price competitive tendering) as a procurement method, do not be surprised if you get adversarial relationships. This might – it should not, but it might – include the withholding of a compliant Regulation 15 (4) health and safety plan until the last possible moment before starting work on site, either to gain negotiating advantage, or prevent advantage being taken.

When the CDM Regulations where first introduced, it was quite common to see massive health and safety plans written in pseudo-legal jargon, which for the most part were both of little use and non-compliant insofar as they did not follow the Appendix 4 format. Since then standards have improved but in the writer's view, they still have some way to go.

Health and safety plans are intended to be of use to those actually carrying out the work on site, and they should be prepared with this in mind. For that reason, and others, it is unwise to write them into the contract. This issue is discussed further in Chapter 9.

The information contained in plans is generally but not solely obtained from risk assessments, either design risk assessments prepared under CDM Regulation 13, or work risk assessments prepared under Management Regulations 3. Preparation of plans takes place over time, but it should add little or no extra cost to the process because plan development takes place utilizing existing systems that should be in place anyway.

Health and safety files: format and content

There is no set format or content for health and safety files, although guidance is given at Appendix 5 of the approved code of practice. Again the development of files takes place over time, and the type of information contained in the file can vary.

The development of files takes place in three distinct and separate but sometimes overlapping phases (see Figure 7.11). During the design stages of projects, the files should contain information of interest to designers, needed to enable them to achieve compliance with CDM Regulation 13. Planning supervisors play a key role in deciding what information is relevant, and in the obtaining of the necessary information through the client. CDM Regulation 11 is cardinal to the process, and requires clients and planning supervisors to ascertain and provide information which may then be placed in the file.

Designing frequently overlaps with the construction stage, and to the extent that it does, the above remains relevant. Also during construction, the files should contain information needed by contractors to enable them to work safely during the building process.

Once the client takes possession of the building or structure, the file should contain information needed in using the building safely, but specifically for any cleaning, maintaining, altering, refurbishing or demolishing work undertaken during its life cycle. The file needs to be reviewed and updated regularly, and specifically in response to any subsequent CDM-compliant construction work.

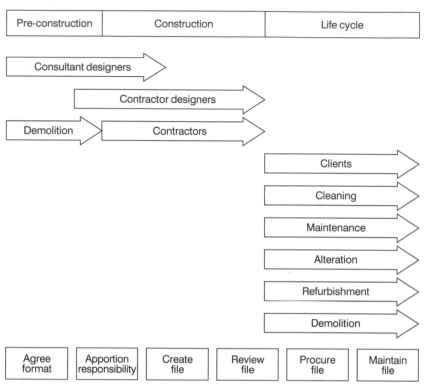

Figure 7.11 File preparation

7.7 Principal contractor's duties

In order to fully understand the role of the principal contractor, it is necessary to consider two sets of regulations, and to bear in mind one fundamental point, which is that a principal contractor must be a contractor, that is, a person or firm which carries out or manages construction as its primary business activity.

The two sets of regulations are:

- **The Management of Health and Safety at Work Regulations 1992;**
- **The Construction (Design and Management) Regulations 1994.**

The management regulations

The Management Regulations are dealt with fully in Chapter 3, but the relevance of the principal contractor role is given at **Regulation 9, Co-operation and Co-ordination**, which requires two or more employers sharing a workplace to co-operate with each other and to co-ordinate their activities. Indeed the approved code of practice accompanying the Management Regulations goes a little further in talking about a main employer or a controlling employer, and offers the guidance that, in the absence of a main or controlling employer, a health and safety co-ordinator should be appointed.

Where two or more employers share a work place, the simplest way of complying with the Management Regulations is for one of them to accept the role of main employer. Most modern construction sites involve two or more employers sharing a workplace, that is, the construction site. Consequently, the role of a principal contractor can be thought of as that of main employer under the Management Regulations, supplemented by regulations 15, 16, 17 and 18 of the CDM Regulations. So, in full the duties are as follows:

Main employer's role

Management Regulations	Duty
3	Risk Assessment
4	Health and Safety Arrangements
5	Health Surveillance
6	Health and Safety Assistance
7	Procedures for Serious and Imminent Danger
8	Information for Employees
9	Co-operation and Co-ordination
10	Persons Working in Host Employers of Self Employed Persons Undertakings
11	Capabilities and Training
12	Employee's Duties
13	Temporary Workers

The above are then added to by the following:

CDM Regulation	Duty
15	Develop at Construction Stage Health and Safety Plan
16	Ensure Contractors Co-operate, Ensure Compliance with the Health and Safety Plan, Give Access to Authorized Persons Only, Display a Form F10 (Revised) On Site, Provide Information for the Planning Supervisor, Make Reasonable Rules in the Health and Safety Plan

| 17 | Ensure Information is Available to Contractors, Ensure Operatives are Adequately Trained |
| 18 | Consider the Views of Operatives on Health and Safety, Co-ordinate the Views of Operatives |

In addition, the principal contractor is given two statutory powers, at Regulation 16, sufficient to enable him to comply with his own obligations in connection with health and safety management. The powers are:

■ to give reasonable directions to contractors;
■ to make reasonable rules in the health and safety plan.

The approved code of practice accompanying Regulation 16 also explains the specific links to other regulations not mentioned above, namely, **Provision and Use of Work Equipment Regulations 1992**.

It is pertinent at this point to remind readers of the six different procurement routes described in Chapter 5, and to mention that the default position is the traditional route, where the main contractor is the principal contractor. However, the following principal contractor appointments could all be made:

■ management contractor is principal contractor;
■ construction manager is principal contractor;
■ client is principal contractor;
■ a subcontractor is principal contractor;
■ some other arrangement is made.

The essential requirements of the appointment being met are that:

■ the principal contractor must be a contractor;
■ the principal contractor must be competent;
■ the principal contractor must be adequately resourced.

The essential difference between these appointments is the way in which organizations and people are employed within the project delivery supply chain.

In the traditional route, the employer is in contract with the main contractor, and he is in contract with the subcontractors. Accordingly the parties have the benefit of a contract matrix which clearly spells out their obligations, and any remedies for breach. Significantly, the main contractor has powers of instruction in the contract.

However, in a construction management or management contracting arrangement, the works package contractors may be in contract with the employer, and the construction manager or management contractor may have limited or even no contractual arrangements with them. Consequently the manager may have no powers in contract to remedy any breach either of contract or of statute. So the CDM Regulations give the manager two statutory powers, sufficient to enable him in his role as principal contractor if appointed, to achieve compliance with his own duties. This is the primary purpose of the mechanism.

Turning now to the principal contractors actual duties, the following points are made.

Develop the construction phase health and safety plan (Regulation 15 (4))

Essentially the principal contractor must set out in the Regulation 15 (4) health and safety plan:

- His responses to the Regulation 15 (1–3) plan. Specifically, he must state the adequate resources he has allocated to dealing with the unusual design risks set out in the pre-tender plan. Regulations 8 and 9 and Guidance Note 36 are relevant to the development of plans.
- The project wide arrangements to be made in connection with his own work methods, and the way in which he intends to do the job.
- Specific risks in connection with specific aspects of the project.
- General risks affecting members of the public, all other people working for other employees at other workplaces affected by work on the principal contractor's site (see Figure 7.10)

The plan does not need to be fully developed before work starts on site. Indeed, it is probably unrealistic and undesirable to develop a plan too far in advance of planned work activity, because many things can change in the interim. However, plans should be developed perhaps six to eight weeks in advance of work programme start dates, and should be continually monitored and reviewed.

Ensure contractors co-operate (Regulation 16)

This is an extension of the management regulation duty upon employers sharing the workplace to co-operate with each other, even when there may be no direct contractual relationship (see Figure 7.12). This is not an easy objective even where there is a contractual relationship. It is even harder

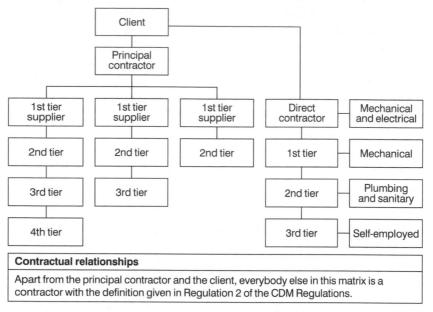

Contractual relationships

Apart from the principal contractor and the client, everybody else in this matrix is a contractor with the definition given in Regulation 2 of the CDM Regulations.

Figure 7.12 Principal contractor – contractor relationships

where there is not. The envisaged means of co-operation are, apart from a contract, the health and safety plan, the reciprocal duty placed on contractors and the statutory powers given to the principal contractor.

The techniques available to principal contractors include

- regular meetings;
- tool box talks;
- the use of Gantt charts, project scheduling programmes, etc.;
- incentive/reward schemes;
- set-off and contra charge mechanisms.

Ensure compliance with the health and safety plan (Regulation 16)

This pre-supposes that contractors are given a copy of the health and safety plan. The CDM Regulations make it clear that not only must contractors be given a copy, but that all operatives working on site must be given a copy too, or at least a copy of those parts of the plan that relate to the work they are doing or are affected by on site.

How the principal contractor does this is a matter for him, but what is envisaged in ensuring co-operation and compliance, which arguably are two sides of the same coin, is that each contractor will carry out his own risk assessments in accordance with Regulation 3 of the Management Regulations, and the principal contractor will supervise and manage this process (see Figure 7.10) using it as the means of developing the Regulation 15 (4) construction phase plan. Consequently both the principal contractor and contractors, however they have may have been appointed or employed, should find themselves doing on site what they had always intended to do anyway.

Compliance should therefore be a matter of choice rather than compulsion, but there are two commonly occurring circumstances where the later may be necessary.

1 Very often, the individual risk assessments of, say two or three contractors engaged in different work activities at the same time on the same workface might reveal only low to medium individual risk exposure. When all three risk assessments are considered collectively by the principal contractor, they may combine to create a high-risk activity. One of the functions of the principal contractor is to identify and deal with precisely this situation.
2 In the model at Figure 7.12, it can be seen that second, third and fourth tier supplier arrangements in construction are matters of routine. In practice, this might work as follows.

The client appoints a direct contractor to install all of the mechanical and electrical services in his new factory. The direct contractor may undertake the electrical services himself, but subcontract the mechanical component to his first tier supplier, another separate company. This company might retain the heating component for itself, but sub contract the plumbing and sanitary work to another company, which then becomes a second tier supplier. But this company might only supply the materials itself and as a matter of routine it might install them using self-employed plumbers (third tier suppliers) employed on lump sum prices.

It would be entirely typical for the direct contractor to know little or nothing about the third tier supplier working for him. He would expect to liaise with his appointed first tier supplier and in the normal course of events would have no direct contract with the second and third tier suppliers.

However, the CDM regulations, first, require this part of the supply chain to co-operate and comply within itself, and second, with the rest of the supply chain shown in the model at Figure 7.12, and third, with the principal contractor.

The principal contractor must make arrangements to ascertain the supply chain links to enable him to know the identities of all the contractors, before he can be in a position to discharge his duties fully.

Authorized persons access: Regulation 15

An authorized person is one who is employed on the site having had project and task specific training in how to do his or her work properly and safely in accordance with relevant health and safety legislation. Consequently, the principal contractor must put in place a system for ensuring everybody working on the job, or visiting in connection with it, is adequately informed and trained to a standard sufficient to comply with this regulation.

This does not mean that the principal contractor has to meet the cost of such a system solely out of his own pocket. There is nothing to stop him implementing an arrangement with subcontractors and suppliers whereby he, the principal contractor, identifies the information and training needs of all persons working on site, provides a means of meeting those needs, and charges his subcontractors and suppliers accordingly. Obviously, any such arrangement should be agreed in advance between the parties.

If a person who is not authorized in this way gains admittance to site, the consequences for the principal contractor can be severe, and apparently unfair. For example, if a ten year old trespasser gains access to site, falls off a scaffold whilst playing on it and breaks his leg, he is not an authorized person and the principal contractor might have a case to answer in explaining how and why he failed to prevent access. Alternatively, one of the self-employed third tier plumbers mentioned above may be properly inducted and trained before starting work on site. He might be ill one day, and, apparently conscientiously, he sends his brother along in his place, who is also a qualified plumber. However, he is not authorized as far as this site is concerned and again the principal contractor might have a case to answer.

Display a form F10 (revised) on site

The construction phase form F10 must contain, inter alia, the addresses and names of the client, principal contractor and planning supervisor, and it must be displayed in a prominent place on site, for two main reasons.

First, to inform people working on the site with the particular interests in mind of lower tier suppliers shown in Figure 7.12. Very often, the third and fourth tier suppliers have no idea of the identity of their ultimate paymaster, or of the person responsible for managing health and safety on site. Form F10 (revised) provides essential information to everybody at work on the site, and to members of the public or other employees affected by the work activity of the site.

Second, any authorized officials, such as an HSE inspector visiting the site, can immediately see who is in charge and can know exactly where to find them.

Provide information for the planning supervisor: Regulation 15

The type of information that the planning supervisor might reasonably require of a principal contractor falls into three broad categories.

First, information relating to the developing design. This might occur where there are significant portions of contractor design work, and the planning supervisor needs, for example, design risk assessments from consultants and contractor designers to continue to develop the Regulation 15 (1–3) health and safety plan, or where there have been significant changes, or as a means of continuing to ensure designers are co-operating with each other and complying with their obligations under Regulation 13.

Second, information needed for the health and safety plan, including work method statements and risk assessments. In other words, information that might be needed to show how contractors are going to work safely, and the principal contractor is going to manage them, on site.

Third, information needed for inclusion in the health and safety file. This might take the form of as-built drawings or operating and maintenance manuals, material data sheets or anything else which in the planning supervisor's reasonable opinion, might be needed in the health and safety file.

Make reasonable rules in the health and safety plan: Regulation 15

The rules the principal contractor can make are those needed to enable him to comply with his owns duties under the regulations, and to ensure that compliance of others under his control, including contractors, designers and authorized persons. Accordingly the health and safety plan might contain rules along the following lines:

> *Designers:* All designers of contractor designer works will carry out design risk assessments using the hazard, risk, control, under nine separate headings, methodology.

> *Contractors:* All contractors shall submit lists of names of all operatives, together with their training records, in a standard format similar to that set out in the company health and safety policy.

> *Authorized persons:* Anyone working on site must possess as a minimum, a valid CITB certificate of competence in health and safety, which they must show the site manager before starting work.

> *Risk assessments:* All contractors will provide written risk assessments and method statements relating to their work activities one week before the programmed start date of their work.

File information: All information requested by the planning supervisor for inclusion in the file shall be prepared and presented in the agreed format no later than seven days after completion of the work to which the information relates. If contractors require access to the file for any reason, the following rules apply.

General rules:
- All power tools shall be 110 V.
- Radios and personal stereos are not permitted
- The site is a no smoking site.
- Site welding is not permitted.
- There shall be no subcontracting of any part of the work without the express written permission of . . .
- Hard hats shall be worn at all times.
- If there is an emergency, here is the evacuation plan . . .

The making of reasonable rules serves as a means not only of instruction, but also of information and communication.

Ensure information and training is available to contractors: Regulation 17

It is envisaged that the means of ensuring information is available to contractors is mainly, but not solely, by way of the health and safety plan and the health and safety file.

The plan is intended for use by operatives, so it should be written in plain English without jargon and with no contractual intent.

File information may, of necessity, be of a technical nature and can therefore use more complex language. It does not necessarily have to be in one place, and could be dealt with by the use of an index, as previously mentioned.

The principal contractor must put into place project specific procedures for distributing the plan, and making available the file. He must also implement procedures for ensuring operatives are adequately trained. Effectively, this requires the following as a minimum:

- a training record procedure;
- an induction procedure.

Training is a major issue in construction at the present time, as it has been throughout the 1990s. The Construction Industry Training Board (CITB) was set up in 1964, and it has a statutory duty to ensure that there is an adequate supply of people trained to appropriate standards to meet the needs of the industry.

The CITB achieves this through close consultation with construction companies, trades federations and representative bodies, and its structure comprises the following:

- the Board itself;
- the training committee;
- the finance committee;
- co-ordinating committee for NVQs;
- federations advisory committee;
- regional advisory committees.

The CITB offers both youth and adult training courses and develops training programmes and materials including workshops, lectures, books and videos together with a range of easy-to-read publications and guidance notes, which go under the title of 'Construction Sites Safety: Safety Notes'. This should form a basic part of any operative training programme.

The CITB offers general training at the level of

- trainees;
- operatives;
- craftsmen;
- site Supervisors;
- site Managers;
- specific site safety issues at all levels.

It is funded by a payroll levy on all firms within the industry above a certain, variable, payroll size, which is currently around £70 000 per annum. Consequently, all but the smallest of contractors should be contributing to the CITB, which is a matter that potential employers may wish to consider before appointing contractors. Do they pay CITB levy and can they demonstrate it?

The CITB has four large training centres at Bircham Newton, in Norfolk, at Erith in Kent, and at Glasgow and Birmingham, as well as seven smaller regional centres. It also employs about 13 safety training advisors whose task is to advise companies in the preparation of staff training within their own organizations.

As well as the CITB, there are also organizations such as RoSPA and the British Safety Council, both of which offer good, reasonably priced training along similar lines to the CITB. Consequently, there is a plentiful supply of good reasonably priced training readily available to principal contractors and contractors, and there really are no excuses for the poor training standards which are regularly encountered in construction.

Consider and co-ordinate the views of operatives: Regulation 18

This is embodied in the old 'suggestion box' approach, but many modern organizations have long since left this behind. It has been replaced with many different types of innovatory schemes such as quality circles or poka yoke techniques (see Chapter 4).

Construction has been notoriously slow in implementing quality systems and for the most part, remains a long way behind current best practice in other industries, particularly manufacturing.

Very often operatives are best placed to see and do something about health and safety shortcomings on site, provided they are given the correct backing and support of supervisors and managers. This regulation is intended to ensure that they do receive both and means that supervisors and managers can be held to account if they fail to provide it.

Summary of the role

The principal contractor role can be thought of as that of main employer, supplemented by a series of duties specific to activities on building sites, taking into account the vagaries of the construction industry and the construction process, which requires the principal contractor to undertake co-ordinating, informing, training, involving and listening activities, and the enforcement of those activities.

7.8 Contractor's duties

Business terms applied to contractors can include the following

- main contractor;
- management contractor;
- construction manager;
- lead contractor;
- head contractor;
- specialist contractor;
- nominated subcontractor;

- named subcontractor;
- approved subcontractor;
- subcontractor;
- labour-only subcontractor;
- self-employed subcontractor.

These can range from very large organizations to very small ones, or even individual persons. They can be employed in two ways:

- through the head contract matrix, in other words, by the main contractor, or his subcontractors;
- through a separate contract matrix, directly with the client.

On small projects, these arrangements are usually fairly straightforward, but on large projects, particularly those involving more than one site, or a number of different workfaces, they can be extremely complex.

Multi-site projects, for example, overhead power lines or underground pipelines, can have a number of main contractors employed at different locations miles apart, and each will have its own separate subcontract network. On the face of it, each site may appear to be a completely separate project, but in fact, the work at each site may contribute to a process which requires co-ordinated action along the length of the power line or pipeline, either on a continuous or on an intermittent basis.

Consequently, the work of subcontractor A at site 1, whilst relatively minor and small scale, may have an impact on subcontractors B, C and D at sites 2, 3 and 4 and yet each may have no idea of the other's existence. The principal contractor/contractor relationship is intended to deal with precisely this type of situation.

The important thing to remember about CDM is that it applies to projects, not contracts, that there can only be one principal contractor at any one time on a project and that there must be a principal contractor appointed from the start of the construction phase until all construction work on the project is completed.

Accordingly, once the principal contractor appointment is made, every other contractor working on the site, however he is employed, is a 'contractor' for the purposes of the CDM Regulations.

The requirements and prohibitions on contractors are set out at Regulation 19 of the CDM Regulations, and in outline are as follows:

- Contractors are, firstly, employers and have the duties given to employers under the HASWA/M HASWA framework. This can be confusing when considering the status of labour-only or self-employed persons, as they can sometimes hold the duties of employers and employee. However, clear guidance is given in the Management Regulations at Regulations 3, 9, 10 and 13.

- Contractors must:
  1 Co-operate with the principal contractor.
  2 Provide the principal contractor with relevant information.
  3 Comply with the principal contractor's reasonable directions.
  4 Comply with any rules in the health and safety plan.
  5 Inform the principal contractor of any notifiable accidents.
  6 Provide the principal contractor with information for the health and safety file.
  7 Provide their own employees with copies of the health and safety plan.

Co-operate with the principal contractor

The construction industry is changing, but it is still dominated at site level by macho, male-orientated, risk-taking, get-on-with-the-job-at-all-costs attitudes and people. Self-employment is the norm at site level, both among operatives and most supervisory and management staff, and it is commonplace to be employed on the basis of lump sum fixed prices.

This brings with it an attitude of self-interest which can be summarized in the view that 'I work for myself you can't tell me what to do, I will do what I have priced for, and to hell with everyone else'. Commercially such attitudes are understandable. In terms of process efficiency they can be disastrous, and in terms of effective health and safety management they are, in the writer's view, a major unrecognized cause of accidents on building sites, because it can become very difficult to implement effective team working when self-interest is encouraged and incentivised.

Nevertheless, the principal contractor must do just that; implement effective team working with two groups of contractors, first those in contract with the principal contractor but not the client, and second those in contract with the client but not the principal contractor.

It would seem a matter of common sense that contractors should want to co-operate with each other, and within the construction process itself. The benefits of such co-operation are, for the most part, demonstrable. Sharing of common resources should and does result in efficiencies and cost savings in certain circumstances. For example, it makes sense for the main contractor to provide a tower crane on many sites, which is then available to all contractors who might need it. The alternative is for them all to make their own hoisting and lifting arrangements. Other shared facilities can include:

■ scaffolding;
■ temporary lighting;
■ temporary power;
■ welfare and office accommodation.

However, it is generally not efficient or cost effective to share the following:

■ small plant and equipment;
■ personal tools;
■ task specific materials;
■ trade specific materials.

Hence the skill of the principal contractor lies in understanding the detailed needs of all of the trades and works package contractors working under him. In so doing he will be able to identify areas of co-operation and potential non-cooperation. That is why the principal contractor must be a contractor. It is axiomatic that problems nearly always arise in areas where non-cooperation is to be found, namely at the interfaces of trade packages, in what Latham (1994) and others have termed 'fuzzy edge disease'. Therein lies the main barrier to cooperation among contractors.

Most of the standard forms of contract and subcontract are rooted in the adversarial system (i.e. JCT and ICE families of contract) and incentivize uncooperative behaviour among subcontractors (contractors).

By way of example, consider three contractors.

Contractor A is employed by the main contractor to cut holes and chases in the structure to enable the mechanical and electrical trades to run their services horizontally and vertically throughout the building.

Contractor B is employed by the main contractor to design and install the mechanical services installation, and must provide contractor A with details of where he needs holes and chases and the sizes required.

Contractor C is employed by the main contractor to do the same as contractor B but with the electrical services installation.

Both mechanical and electrical services are to be fitted into the same service zones at the same time, they must be commissioned and tested together, and they must use the same holes and chases.

Suppose that contractor C is late giving details of his required holes and chases to contractor A, who in turn, is late in constructing them, which therefore delays contractor B. These delays may or may not constitute breach of contract.

All three contractors might incur loss and expense because of these delays (breaches) and so might the main contractor, who in this case is the employer of all three. This could set up a complicated chain of claims and counter claims as each party seeks to minimize the damage to itself flowing from the alleged breaches. Contractors A and B would almost certainly have a claim arising from contractor C's activities, but they are not in contract with him, whereas they are with the main contractor. So they would have to claim under the relevant clauses of their contract with him, and he in turn might have to pay their claims in terms of time and money, even though he may not be to blame. He would then be obliged to seek reimbursement from contractor C by way of set-off or claim. This is a simple and simplistic example. Things can get much worse when considering issues of function, performance, technical standards and quality.

The point is this. In order to be properly paid in time and money, contractors A and B will almost certainly be required by their forms of subcontract to issue written notices specifying:

- each and every breach of contract;
- the reasons why breaches occurred;
- who caused the breach;
- the effects of each and every breach;
- particularized loss and expense flowing from the breach.

This is an inherently adversarial system which even in the simple example quoted above, usually results in an avalanche of notices and a storm of correspondence as each party seeks to establish and preserve its entitlements under the contract, because the alternative is, that in the absence of appropriate notices and correspondence, all entitlements may be lost.

Amidst this mayhem, cooperation is often the last thing on everyone's mind and health and safety management arrangements fly out of the window.

There are alternatives. The GC Works Series and the Engineering Construction Contract families of contract are said to be cooperative systems which encourage the effective implementation of team working through early warning meetings, leading to the identification of compensation events, which are pre-priced and agreed in terms of time and money.

The continuing importance of the correct choice of procurement route and contract strategy can be seen at this level.

Most experienced commentators agree that adversarial, unfair contract terms are a major barrier to cooperation within the supply chain (Latham, 1994. Cox and Thompson 'Contracting for Business Success', 1997) and any prudent main contractor acting as principal contractor will avoid them.

Provide the principal contractor with relevant information

The regulations specifically mentions risk assessments made under Management Regulation 3, and it is safe to assume that the primary purpose of this particular regulation is to prompt, indeed compel, principal contractors into ensuring there is a rigorous risk assessment procedure in place.

In other words, the principal contractor should require all contractors to submit risk assessments to him, complying with Management Regulation 3, of all work activity on site before it starts.

However, the wording of the regulation is broad enough to clearly include any information, not just risk assessments, which might be relevant to health and safety of

- persons at work carrying out construction,
- persons affected by the work of such a person, or
- which might justify a review of the health and safety plan.

This might include, *inter alia*, the contractor's health and safety policy, or any old drawings or surveys, or specialized local knowledge that contractors might have.

Comply with the principal contractor's reasonable directions

Specifically, those given under Regulation 16 (2)A. This is the reciprocal of the principal contractor's statutory power of instruction, and is the other half of the mechanism. It comes into its own where the principal contractor has no contractual relationship with the contractor. It is no good any direct contractor of a client refusing to except the principal contractor's instruction issued under 16 (2)A on the grounds that there is no contract. The principal contractor's statutory power is good enough. The only grounds for objection is whether or not the principal contractor's instruction is needed and reasonable to ensure his own or the contractor's compliance with their respective obligations.

Similarly the power is useful for dealing with the subcontractor of a subcontractor, for the same reasons, that the main contractor may not be in contract with the sub-subcontractor.

Comply with any applicable rules in the health and safety plan

Project specific rules can be made by:

- clients;
- planning supervisors;
- principal contractors.

Clients may make rules under headings 2 to 7 of Appendix 4 of the approved code of practice, namely:

2 existing environment
3 existing drawings
4 the design
5 construction materials
6 site wide elements
7 overlap with client's undertaking.

It makes sense for clients to be given this facility for they are more likely than anyone else to possess information on headings 2, 6 and 7, particularly where they already occupy the site or buildings on the site.

Planning supervisors may also make rules under the same headings, and they should be best placed to do so under headings 3, 4 and 5, using information obtained from designers, and to assist clients with appropriate advice relating to headings 2, 6 and 7.

Principal contractors may make rules under the above headings if they so choose, but are more likely to make them under headings 8 and 9.

8 site rules
9 continuing liaison

Most of the rules to be made are simply matters of common sense and good business practice, and consist of information that is simple and readily available. Unfortunately, all too often the people in possession of this information take it for granted that others know these things when they do not.

The writer has walked around numerous business enterprises with many different clients, all of whom are mines of information about their premises but have great difficulty in finding written or drawn information that I have needed to enable me to provide them with the service that they require. For example, clients rarely know how many service entries (gas, water, electricity, telephones) they have into their buildings or their locations. Many are vague on whether they have single-phase or three-phase power, and things such as distribution boards or spare ways can be a foreign language.

The criteria for writing rules in health and safety plans must always be relevance, simplicity and reasonableness to the extent needed to ensure the compliance of the principal contractor and his contractors.

Contractors have a statutory duty to comply with any such rules, whether or not they are in contract with the principal contractor.

Inform the principal contractor of any notifiable accidents

This is an extension on all employers of the duties set out in **RIDDOR 1996**.

Provide the principal contractor with information for the health and safety file

There are three aspects to this duty, relating to information, namely:

- information in the principal contractor's possession;
- information he could ascertain through reasonable enquiry;
- information that the planning supervisor might reasonably require.

In practical terms, the type of information that will be in the possession of contractors will most likely comprise as-built drawings and records of the work that they have actually done on site.

The main question that then arises is, to what extent is this information relevant for inclusion in the health and safety file? For example, there is little point in asking the plastering subcontractor to provide detailed records or as-built drawings of the work that he has done.

This issue, more than any other, is best dealt with thoroughly in contract. For the avoidance of doubt, and to enable contractors to allocate adequate resources to the task, and to price accordingly, it is sensible to spell out in detail in the tender documentation exactly what information contractors would be expected to contribute to the health and safety file.

Contractors can therefore expect to see the file information requirements clearly set out for them in tender enquiries, or in contract documentation. However, if this is not the case, then contractors must still provide it if and when requested. However, they may be able to request payment in the form of extra time and money both under an adversarial contract, and a cooperative contract.

Provide employees with copies of the health and safety plan

The whole point of health and safety plans is that they should find their way into the hands of the people who are actually carrying out the work on site. Thus each employer, whether using directly employed or self-employed operatives, must ensure that their employees are provided with a copy of the plan, or at the very least, those parts that are relevant to work which the operative is doing. The minimum information to be provided is:

- the name of the planning supervisor;
- the name of the principal contractor;
- the relevant parts of the health and safety plan relating to work to be carried out by the person concerned.

Thus contractors must make a judgement on this latter point. It is also important to remember that from time to time the plan may be reviewed and updated, and that operatives should be furnished with current copies of the plan.

The Health and Safety Executive publish useful practical guidance on this topic entitled 'Managing Contractors: A Guide For Employers' ISBN 8 7176 1196 5.

7.9 Chapter summary and conclusion

This very long chapter has examined the major provisions of the CDM Regulations, taking an in-depth look at the five new dutyholders created by the regulations, their roles and the two statutory documents created to assist in achieving compliance.

The main impact of the CDM regulations is to introduce the hierarchy of risk control into the arrangements made for the design and management of construction projects.

Health and safety plans are intended to assist with implementing health and safety management arrangements during the design and construction phases of projects. Health and safety files are intended to do the same over the life cycle of any building or structure.

Clients are expected to play a much more active role in the arrangements made to design and manage projects, although HSE have made it clear that they expect all dutyholders to achieve far better standards, which will increase in time and which they say will require a radical change in culture throughout the construction process.

The importance of the procurement route and contract strategy on health and safety arrangements cannot be understated.

Adversarial forms of contract incentivize uncooperative behaviour, which in turn leads to lack of cooperation on site. This is one of the major unrecognized causes of unsafe work environments, and unsafe work systems, and is an area where further research is urgently required.

Prevailing cultures within the construction process are a serious barrier to improving health and safety standards. Vested interest remains one of the main drivers of the process, along with the continuing separation of designing from constructing, and the fragmentation of the supply chain. These issues diminish the ability of project supply chains to work consistently together as teams, and in consequence, increases the potential for unsafe and unhealthy work practices, not only on site, but also in the designing process.

8

The Construction Regulations 1996: practical application

8.1 Introduction and background

The full title of the so-called new Construction Regulations is the Construction (Health, Safety and Welfare) Regulations 1996. They were laid before Parliament on 27 June 1996 and they came into force on 2 September 1996. The regulations were made by the Health and Safety Commission under the Health and Safety at Work Act etc. 1974, and represent the UK's implementation of Annexe IV of the Temporary and Mobile Construction Sites Directive (TMCSD).

The new regulations apply to all construction work all of the time, unlike CDM which sets minimum thresholds for notification and application (see Chapter 7).

The new regulations replace three of the old four sets of construction regulations made in the 1960s, namely:

■ The Construction (General Provisions) Regulations 1961;
■ The Construction (Workplaces) Regulations 1966;
■ The Construction (Health and Welfare) Regulations 1966;

The fourth of the old set, the Construction (Lifting Operations) Regulations 1961, have also been repealed and replaced by the Ligting Operation and Lifting Equipment Regulations (LOLER) 1998 – see Chapter 10.

The new regulations comprise 35 new or revised provisions in one main set, in place of more than 90 old regulations distributed through three sets. They make it easier to understand and comply with the law by clarifying and simplifying the new provisions whilst bringing them up to date with modern procurement and contracting customs and practice.

With the introduction of the new regulations, the management framework for health and safety in construction in the UK is now complete. The clue is in the titles. The CDM Regulations apply to those who design and manage construction. The new Construction Regulations 1996 apply to those who carry out construction work on site. Between the two sets of regulations, minimum standards are set throughout the life cycle of construction projects. There is no overlap or conflict between the two sets.

8.2 Why is there no ACoP with the new regulations?

HSC say that, having taken the advice of various bodies including CONIAC, CBI, TUC and others, they have not altered the provisions of the old regulations enough to warrant an ACoP (see Chapter 1, Section 1.4).

HSC say that all they have done is modernize the old laws, consolidated them into a single set of provisions and introduced new provisions only where necessary to comply with Annexe IV of TMCSD where they were not reflected in the old legislation. Accordingly, it was felt that the significant provisions of the new regulations are already well known and widely established, and that the new requirements are not sufficiently complex to justify an ACoP.

8.3 How do the regulations work?

First, the new Construction Regulations use the same definition of **construction work** as used in the CDM Regulations. This ensures that the two sets exist side by side, with no overlaps and no, as yet, apparent gaps or omissions from coverage (see Regulation 2 of both CDM and Construction Regulations).

Second, until now there has never a statutory definition of **construction site**. However, this is now dealt with at Regulation 2 of the Construction Regulations, where a definition is made. However, it is not a prescriptive definition, it does not tell dutyholders what to do, instead it tells them how to achieve compliance. This effectively propels the dutyholder into defining the site boundaries on an on-going basis. Thus, CDM applies to projects, and the Construction Regulations apply to any site or sites within a project. For example a power distribution project may comprise ten different sites at ten different locations between A and B which are ten miles apart. The project is to distribute power across the ten miles distance from A to B. CDM applies to the project. The Construction Regulations apply to each site, whenever work is in hand on that site.

Third, the regulations apply to:

- employers,
- the self employed, and
- employees
- persons who control the way. . . .

This is an apparent departure from the approach adopted in the CDM Regulations, where five new dutyholders were created, and the term contractor was used in connection with carrying out construction work on site.

There is a very good reason for this departure. It is not just contractors who carry out or manage construction work on site. Many people and organizations do it, not as their main business activity, but as a necessary part of that activity. Examples include property developers, housing associations, local authorities, NHS trusts, factory owners and facilities managers. Accordingly, if duty holders had been limited to contractors, then all those who are not contractors, but who regularly carry out construction, would have been outside the provisions of the new regulations.

As has been made clear throughout this book, the terms **employer** and **self-employed** apply to everybody in equal measure. A factory owner who undertakes construction work has exactly the same responsibilities under the Construction Regulations 1996 as a contractor undertaking the same work; they are both employers. The only way the factory owner can transfer responsibility under the Construction Regulations is by employing the main contractor.

Fourth, an interesting feature of the regulations is that some of them apply not just to employers, employees and the self-employed, but also to **persons who control the way in which construction is carried out** (see in particular Regulation 22, welfare facilities).

8.4 Case study

A multinational chemical manufacturer entered into a turnkey contract with Powergen Projects and Powergen CHP for a new combined heat and power plant at a site in Northwich, Cheshire. The work involved the design, engineering, procurement and construction of a new plant at a cost of about £350 million, and included not only the heat and power generating plant but also overhead transmission and distribution systems to a series of sites both within the 100 Ha complex and outside it, in some cases up to 12 km away.

The project was broken up into six work packages, mainly based on geographical considerations, and upon identifiable separate processes.

Sites 1, 2 and 3 were all standalone parts of the CHP facility at different locations within the existing complex. Each work package at each site was awarded to a different works package contractor who was the employer for that site.

Sites 3, 4 and 5 were much more difficult to deal with. First, they were all outside the main complex. Second, they were active at different times, and third, they were a long way away from each other and from the complex itself. Nevertheless, a logical and appropriate packaging arrangement was made and a main contractor was appointed to each site, and he was the employer for that site.

At sites 3, 4 and 5 each employer was responsible for implementing all of the Construction Regulations, including Regulation 22, welfare facilities.

At sites 1, 2 and 3 each employer implemented all of the Construction Regulations except Regulation 22, welfare facilities. The chemical manufacturer took on compliance with this regulation himself, mainly because he had a large semi-permanent contractors' compound within the complex for the

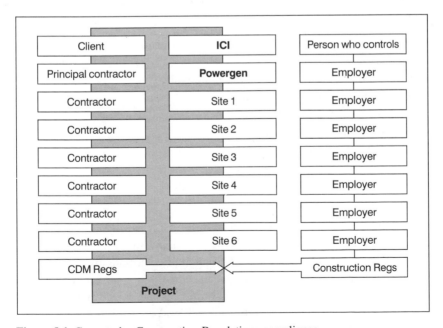

Figure 8.1 Case study: Construction Regulations compliance

use of the many external contractors constantly employed on various projects, as well as this one, at the works.

So there were six employers at six separate sites, with the client (the chemical manufacturer) undertaking the duties of the person who controls the way, as set out as at Regulation 22, that dealt with Construction Regulation compliance. Together the six sites were part of one project, the CHP plant, and the principal contractor appointed to the project in accordance with Regulation 6 of the CDM Regulations was Powergen, who bore the statutory responsibilities of the principal contractor but in fact delegated the tasks to the six employers and the clients, and then ensured that they were carried out to the required standard (see Figure 8.1).

8.5 Costs and benefits

HSC/HSE estimate that costs to industry of the new regulations in Year 1 (1996–97) were £45–50 million. After that, further costs of about £6 million per annum are expected but could rise to £19 million depending upon cyclical training and workload factors. For 25 years costs of about £113–130 million at net present value are forecast, against current industry turnover levels of about £50 billion per annum. The new regulations will therefore cost about 0.2 per cent of turnover.

The main areas of extra cost arising from the proposals in Annexe IV of TMCSD are:

- Intermediate guard-rail (or equivalent) sufficient to prevent falls from height;
- Vehicle/pedestrian segregation on building sites;
- Additional minimum welfare facilities;
- Additional inspections of scaffolding;
- Additional inspections of fire egress;
- Additional inspections of emergency egress;
- Safety features for doors and gates;
- Protection from adverse weather conditions;
- Emergency lighting provisions.

About 60 per cent of accidents in the construction industry involve people falling from heights greater than 2 metres or being struck by vehicles. It is estimated that the extra protection measures will contribute towards a 10–15 per cent decline in these types of accidents, equating to an overall decline in all accidents on sites of 6–9 per cent. From this the annual benefits to the whole of society are put at £18 to £20 million per annum, or somewhere around £160 million net present value over 25 years.

8.6 Links to other regulations

CDM Regulations

As previously mentioned, the same definition of construction work and structure is used in both the CDM and the Construction Regulations.

Work equipment

There are links to the Provision Use of Work Equipment Regulations 1992.

Lifting operations

The old 1966 lifting operations regulations remain in force at time of publication (1999), but it is expected that the scope of lifting operations will be modified to come into line with the definition of construction work to ensure consistency. The legislators are also likely to take the opportunity to modernize the regulations to take account of, for example, new types of crane.

Workplace (Health Safety and Welfare) Regulations 1992

The new regulations require employers and/or persons who control the way in which construction is carried out, to achieve welfare and environmental standards on building sites analagous to those established by the workplace regulations for permanent workplaces.

Management of Health and Safety at Work Regulations 1992

There are specific links relating to information and training, and to the risk assessment requirements of the Management Regulations.

8.7 Arrangement of the regulations

There are 35 regulations and 10 schedules. A few of the regulations are informative:

Regulation 1: Citation and commencement
Regulation 2: Interpretation
Regulation 3: Application
Regulation 4: Persons upon whom duties are placed by these regulations.

A few of the regulations are procedural:

Regulation 31: Exemption certificates
Regulation 32: Extension outside Great Britain
Regulation 33: Enforcement in respect of fire
Regulation 34: Modifications
Regulation 35: Revocations.

Thus Regulations 5 to 31 inclusive can be thought of as **working regulations** which apply to work activities on **construction sites**.

The working regulations interact in the following way. Consider Regulation 6(3). This regulation requires dutyholders to take suitable and sufficient steps to prevent, so far as is reasonably practicable, any person falling. It goes on to require the provision and use of suitable and sufficient guard-rails and toeboards, barriers or other similar means of protection, which must comply with Schedule 1 – a very general description which many people would struggle to comprehend.

Schedule 1 then adds further important information including:

- Main guard-rail to be a minimum height of 910 mm above falling edge.
- Unprotected gaps (i.e. between guard-rail and toeboard) must not exceed 470 mm.
- Toeboards to be not less than 150 mm high.

Thus it is not difficult to calculate that, with a minimum height to top of toeboard of 150 mm, and a minimum height from falling edge to hand rail of 910 mm, there is an unprotected gap of 760 mm.

Therefore, at least one horizontal intermediate guard-rail must be provided which, if placed on the horizontal centre line between the hand-rail and the top of the toeboard will reduce the unprotected gap to 380 mm, thus complying with the requirements of Schedule 1. However, it can also be easily seen how more than one intermediate guard-rail might be needed. An important point to remember is that the regulations are intended to be read in conjunction with the appropriate HSE guidance sheet, which in this case might include:

- Safe use of ladders: construction sheet no. 2
- External access scaffolds: construction sheet no. 3
- Safety in roof work
- Tower scaffold: construction sheet no. 10.

There may also be relevant British Standards to be taken into account, together with the Construction Industry Research Information Association (CIRIA) guides, which may be relevant.

The new Regulations are basically goal setting, but the schedules retain some of the prescriptive characteristics of the old regulations, and also demonstrate why an approved code of practice is not thought necessary.

The standard terminology in use in health and safety law is used throughout the regulations including terms such as:

- ensure
- suitable and sufficient
- practicable
- so far as is practicable
- so far as is reasonably practicable
- competent person
- of suitable design and construction
- of sufficient strength and capacity.

The usual three levels of duty are present, namely:

- absolute
- practicable
- Reasonably practicable.

Turning now to the Regulations themselves. They are summarized as follows.

8.8 Summary of the regulations

Regulation 1: Citation and commencement

The regulations are the Construction (Health Safety and Welfare) Regulations 1996. They came into force on 2 September 1996.

Regulation 2: Interpretation

The following is a summary of some of the more important provisions under this heading.

- **Construction site** means any place where the principal work activity being carried out is construction work.
- **Construction work** includes any building, civil engineering or engineering construction work and sets out five separate categories (see also CDM Regulations).

- **Place of work** means any place which is used by any person at work for the purposes of construction work or for the purposes of any activity arising out of or in connection with construction work.
- **Plant and equipment** includes any machinery, apparatus, appliance or other similar device or any part thereof used for the purposes of construction work and any vehicle being used for such purpose.
- **Structure** re-uses the definition given in the CDM Regulations.
- **Traffic route** means any route, the purpose of which is to permit the access to or egress from any part of a construction site for any pedestrians or vehicles or both, and includes any doorway, gateway, loading bay or ramp.
- **Vehicle** includes any mobile plant and locomotive and any vehicle towed by another vehicle.
- **Working platform** means any platform used as a place of work or as a means of access to or egress from that place and includes any scaffold, suspended scaffold, cradle, mobile platform, trestle, gangway, run, gantry, stairway and crawling ladder.

Note: For the sake of clarity readers are reminded that the definition of construction work is very broad and applies to conversion, fitting out, commissioning, renovation, repair, upkeep, redecoration, maintenance, site clearance, exploration and investigation, decommissioning, demolition or dismantling, and services installations.

Regulation 3: Application

Regulation 3 applies to construction work carried out by persons at work. They do not apply to any workplace on a construction site which is set aside for purposes other than construction work (this is because other regulations might apply to that work activity in that workplace).

The following regulations apply only to construction work carried out by persons at work at a construction site.

15   Traffic
19   Emergency routes and exits
20   Emergency procedures
21   Fire detection and fire fighting
22   Welfare facilities
26   Good order

Regulation 4: Persons upon whom duties are imposed by the regulations

The duties under these regulations fall on:

- employers
- self-employed
- employees.

The duties also relate to any person who controls the way in which any construction work is carried out by a person at work, insofar as they relate to matters which are within his control. Important exceptions are made in relation to Regulation 22 (welfare facilities) and Regulation 29 (inspection) which expressly say on whom the duties are imposed.

Regulation 5: Safe places of work

This regulation requires every place of work to be made and kept safe for any person working there and this includes the following:

- Provide and maintain safe access.
- Provide and maintain safe egress.
- Provide adequate working space.
- Workplace must be suitable.
- Exclude access to any unsafe place.

The above does not apply whilst the place of work is being made safe, provided steps are taken to ensure the safety of persons involved in that work.

Regulation 6: Falls

This regulation needs to be read in conjunction with Schedules 1 and 2. The regulation sets out the duties and the dutyholders, and the schedules set out the minimum steps the dutyholders might need to take in order to comply.

Duties

Most of the duties are qualified by the terms *suitable* and *sufficient* and *so far as is reasonably practicable*.

The duties are:

- Take steps to prevent persons falling.
- Take steps to prevent persons falling where there is a risk of falls from a height of greater than 2 metres, at access or egress and at the workface.

Means of prevention of falls

The regulations envisaged that the following means of fall prevention may be used:

- guard-rails
- toeboards
- barriers
- other similar means
- working platforms
- personal suspension equipment
- means for arresting the fall
- ladders
- scaffolds.

Nature of the work

The regulations make distinction in the work being done as follows:

- the nature of the work
- duration of the work
- the way in which work is being done
- the way in which equipment is being used
- circumstances where it is not reasonably practicable.

Turning now to the interaction of Regulation 6 with Schedules 1 and 2, when they are read in conjunction, the following emerges:

- Provide guard-rails, toeboards and barriers complying with Schedule 1.
- If work platforms are used, provide a sufficient number, complying with Schedule 2.
- On work of short duration where it may not be reasonably practicable to achieve full compliance then it is permissible to provide personal suspension equipment complying with Regulation 3.
- If none of the above are possible, provide fall arrest devices and comply with whatever is possible. Do not remove the fall arrest devices for any reason other than for the purpose of the movement of materials.
- Do not use ladders unless essential.
- If it is necessary to use ladders they must comply with Schedule 5, and certain parts of Regulation 6 do not then apply. For example, it is not necessary to use personal suspension equipment or fall arrest devices.

If ladders are to be used, they must be used in the following way:

■ They must be of suitable and sufficient strength.
■ There must be no displacement.
■ Secure all ladders longer than 3 metres if possible.
■ If it is not possible, station a person at the foot.
■ Prevent ladders from slipping and falling.
■ Provide sufficient hand-hold at the top or:
    – Extend ladder past access level by sufficient height;
    – If longer than 9 metres, provide safe landing or platform.

When guard-rails are used:

■ They must be of suitable and sufficient strength and rigidity.
■ There must be no displacement.
■ Supporting structure must be adequate.
■ Main guard-rail to be a minimum of 910 mm above falling edge.
■ Unprotected gap not to exceed 470 mm.
■ Toeboards to be a minimum 150 mm high.
■ They must prevent falls of persons, materials and objects.

When working platforms are used, consider any supporting structure to be used in conjunction with the platform. Supporting structure is defined as 'any structure used for the purpose of supporting the working platform' and includes any plant and equipment used for that purpose.

The Schedule then goes on to set out requirements for working platforms under the following headings:

1 Condition of surfaces.
2 Stability of supporting structure.
3 Stability of working platform.
4 Safety on working platforms.
5 Loading.

The various requirements are summarized as follows:

■ Surfaces of supporting structures must be stable, suitable and strong enough to support the platform and its load.
■ The supporting structure is to be strong, rigid and stable.
■ The platform is to be securely attached to the structure.
■ The platform to be stable when altered or modified.

- The platform must not be displaced.
- The platform must be taken down without displacement.
- The platform must be big enough to permit safe work passage.
- The platform must be big enough to permit safe use.
- The platform must provide a safe working area.
- The platform must be not less than 600 mm wide.
- There must be no gaps in the work platform surface.
- There must be no slipping, tripping or trapping hazards on the platform.
- Provide adequate hand-holds and foot-holds.
- Adequately maintain the platform.
- Load the platform safely.

Applying the schedules to the regulations

Generally two conditions must be considered:

- access and egress to the workplace;
- at the workplace.

1 Access and egress

The key question here is, is any person liable to fall a distance of more than 2 metres? If so, take suitable and sufficient steps to prevent any person falling.

Generally, examination of the regulations reveals that they encourage the use of properly constructed scaffold stairways and the like, instead of ladders, as a means of access and egress to workplaces and that is one of the many conditions which is covered by this part of the regulation. Where this is done, that is a scaffold stairway is used instead of a ladder, then all of the above apply but there is no need to use toeboards.

2 At the workplace

Again the question to be asked, is the person(s) at work liable to fall a distance of 2 metres or more at the workface? If so, then take suitable and sufficient steps to prevent persons falling including:

- Comply with access and egress requirements.
- Provide sufficient working platforms: Schedule 2.0.
- Condition of surface: Schedule 2.2.

- Stability of supporting structure: Schedule 2.3.
- Stability of work platform: Schedule 2.4.
- Safety on working platforms: Schedule 2.5.
- Loading: Schedule 2.6.

If all of the above cannot be met either because the work is of short duration or because of the nature of the work: then comply with what is possible, but in any event provide personal suspension equipment. This must comply with Schedule 3:

- suitable and sufficient strength;
- consider load and work to be done;
- securely attached to structure;
- means of attachment to be suitable;
- sufficient strength and stability of attachment;
- prevent people slipping from PSE;
- install to prevent uncontrolled movements.

Where this cannot be achieved or the working platform/guard-rail requirements cannot be met because of the nature of the work or the short duration, then provide suitable and sufficient means for arresting the fall of any person. This must comply with Schedule 4:

- any equipment includes nets and harnesses;
- suitable and sufficient strength;
- securely attached to structure or plant;
- structure or plant to be strong and stable;
- equipment itself must not injure.

When moving materials, guard-rail, toeboard, barriers and fall arrest devices may be removed to the extent necessary to do the job in hand but must be replaced as soon as practicable. (Note that the requirements relating to working platforms and personal suspension devices are not treated in this manner – users can hardly temporarily take off an abseil harness halfway up a chimney.)

When ladders are in use, try to avoid using them as a means of access and egress but if this cannot be avoided, consider the following:

- nature of the work;
- duration of the work;
- Risks to safety of any person.

If it is decided to use ladders, two conditions appertain:

■ Regulation 6(3) does not apply;
■ Schedule 5 does apply.

All equipment referred to in this regulation must be properly maintained.

Scaffolding must be installed, altered and dismantled under the supervision of a competent person.

The installation and erection of personal suspension equipment and fall arrest devices must be carried out under the supervision of a competent person. Installation does not include the personal attachment to the equipment; users must be competent to do that themselves.

If a scaffold stairway, including rest platforms, has been built for use solely as a means of access and egress to a workplace, then toeboards are not required provided no materials or substances are stored on it.

Generally this regulation applies to persons carrying out the work, either the permanent works or the temporary work necessary to prevent falls, but it can also apply to any person. This can and probably does extend not only to persons with legitimate business on the site, for example the architect, the consulting engineer, visiting member of the contractor's staff, building control officers and so on, all of whom would be authorized persons within the definitions set out in the CDM Regulations, but would probably also include unauthorized persons – that is those with no legitimate business on the site, and who may even be trespassers.

At first glance this Regulation may appear to be both complex and lengthy, but this is not the case. More than anything, the Regulation is modernized, taking into account the widespread use of safety harnesses, abseil equipment, lightweight aluminium scaffold towers and the like in place of tubular access scaffolding, ladders and bosun's chairs.

Regulation 7: Fragile materials

Suitable and sufficient steps must be taken to prevent any person falling through any fragile materials from a height greater than 2 metres. The suitable and sufficient steps can include:

■ platforms
■ coverings
■ similar means
■ guard-rails
■ warning notices.

Suitable and sufficient steps must be taken to support the weight of any person passing or working from any of the above, or near any of the above.

Regulation 8: Falling objects

Take suitable and sufficient steps to prevent danger to any person from and to prevent falls of materials or objects. Include the provision of:

- guard-rails, toeboards or barriers;
- working platforms, complying with Schedules 1 and 2.

If full compliance is not possible, then take steps to prevent persons being struck.

- No tipping from height where it might cause injury.
- Store materials to prevent danger from collapse, etc.

Regulation 9: Stability of structures

Take all practicable steps to prevent danger to any person working on new or existing structures resulting from instability or temporary state of weakness leading to accidental collapse.

Load structures safely and ensure that buttresses, temporary supports or temporary structures are erected or dismantled under the supervision of a competent person.

Regulation 10: Demolition and dismantling

Take suitable and sufficient steps to ensure that any demolition or dismantling, where there is a risk of danger to any person, is carried out in such a way as to prevent danger and is supervised by a competent person.

Regulation 11: Explosives

Explosive charges can only be used if suitable and sufficient steps have been taken to ensure that no person is exposed to risk of injury from the explosion or from protected or flying material from the explosion.

Regulation 12: Excavations

Take all practicable steps:

- to prevent danger to any person;
- in new, existing or parts of excavations;

- in a temporary state of weakness or instability;
- whilst carrying out construction work;
- prevent accidental collapse.

Prevent any person being buried or trapped by falls or dislodgements.

Excavation to be sufficiently supported as early as practicable using suitable and sufficient equipment, and supervised by a competent person.

Prevent persons, plant or equipment or accumulations of earth falling into any excavation.

No material, vehicle, plant or equipment placement or movement should take place near an excavation where it is likely to cause collapse.

Take suitable and sufficient steps to prevent any risk of injury from underground cables or services. Do not attempt to carry out excavation work otherwise.

Regulation 13: Cofferdams and caissons

Every cofferdam or caisson should be of suitable design or construction, of suitable and sound material and of suitable strength and capacity for the purpose for which it is used and shall be properly maintained.

The construction, installation, alteration or dismantling of a cofferdam or caisson shall take place under the supervision of a competent person.

Regulation 14: Prevention of drowning

During the course of construction, take suitable and sufficient steps where any person is liable to fall into water or liquid at the risk of drowning.

- prevent falling;
- minimize risk of drowning;
- ensure rescue equipment is provided;
- ensure rescue equipment is maintained;
- ensure prompt rescue;
- ensure safe transport over water;
- provide suitably constructed vessel;
- ensure proper maintenance;
- vessel to be controlled by competent person;
- no overcrowding or overloading.

Regulation 15: Traffic routes

Every construction site shall be so organized that pedestrians and vehicles can move safely and without risks to health.

Traffic routes shall be suitable for the persons or vehicles using them, sufficient in number, in suitable positions and of suitable size.

The use of traffic routes shall cause no danger to:

- health and safety of pedestrians;
- safety of vehicles;
- persons near the route;

Doors or gates to be separated from traffic route.

Pedestrians must be able to see approaching vehicles or plant.

Ensure safety by means of sufficient separation between vehicles and pedestrians. If not possible, provide other means for protection of pedestrians.

Provide effective arrangements to warn against crushing or trapping.

One exit point per loading bay for exclusive pedestrian use.

Provide pedestrian doors next to vehicle gates.

Mark door and cause no obstruction.

Do not drive vehicles on traffic routes unless the route is free from obstruction and permits sufficient clearance. If not possible, provide warning of obstruction or lack of clearance.

Every traffic route shall be indicated by suitable signs where necessary for reasons of health or safety.

Regulation 16: Doors and gates

- Fit doors, gates or hatches with suitable safety devices.
- Retain doors in tracks.
- Retain up-and-over doors.
- Prevent trapping on power assisted doors.
- Provide manual/automatic over-ride to power doors.

This regulation does not apply to doors, gates or hatches forming part of any mobile plant and equipment.

Regulation 17: Vehicles

Suitable and sufficient steps should be taken to prevent or control the unintended movement of any vehicle.

Suitable and sufficient steps should be taken to ensure the person having effective control of a vehicle gives warning to any person who is at risk from the movement of the vehicle.

Drive, operate or tow in a safe manner.

Load for safe driving operation or towing.

No riding unless in a safe purpose-made place.

No loading or unloading of loose material unless a safe place is provided and maintained.

Muckshift vehicles to be prevented from falling or overturning into excavation pits, water embankments or earthworks.

Suitable plant and equipment to be provided and used for replacements of de-railed rail vehicles.

Regulation 18: Prevention of risk from fire, etc.

Suitable and sufficient steps should be taken to prevent the risk of injury to any person during the course of carrying out construction work arising from:

- fire or explosion;
- flooding;
- any substance liable to cause asphyxiation.

Regulation 19: Emergency routes and exits

Where in the interests of the health and safety of any person on a construction site, a sufficient number of suitable emergency routes and exits shall be provided to enable any person to reach a place of safety quickly in the event of danger.

- Route to lead directly to identified safe area.
- Route to be kept clear and free from obstruction.
- Provide emergency lighting on the route.
- Use route or exit at any time.
- Indicate route by suitable signs.

Regulation 20: Emergency procedures

Implement suitable and sufficient procedures for dealing with foreseeable emergencies.

- Include any necessary evacuation procedures.
- Ensure possible users are familiar with procedures.
- Arrange regular tests.

Regulation 21: Fire detection and fire fighting

Provide, locate and maintain the following:

- suitable and sufficient fire fighting equipment;
- suitable and sufficient fire detectors and alarm systems;
- examine and test;
- ensure equipment is easily accessible;
- provide instruction and training;
- provide adequate signs;
- apply risk control measures.

Regulation 22: Welfare facilities

Employers and the **self-employed** attract the specific duties of complying with this regulation. However, **any person in control of a construction site** has to ensure compliance. Under most of the standard forms of contract this means that the employer must ensure that the main contractor complies with **paragraphs 3–8 of the regulation**, which must be read in conjunction with **Schedule 6**. Collectively the following **minimum requirements** must be met on **all construction sites regardless of size**.

Sanitary conveniences

- Sanitary conveniences shall be provided.
- They must be adequately ventilated and lit.
- They must be kept clean and orderly.
- Separate facilities for men and women must be provided.

Washing facilities

- Washing facilities shall be provided.
- Locate near every sanitary convenience.
- Locate near changing rooms.
- Provide clean hot and cold water.
- Provide soap.
- Provide towels.
- Ensure sufficient ventilation.
- Ensure sufficient lighting.
- Keep clean and orderly.
- Provide separate facilities for men and women.

Drinking water

- Drinking water should be provided.
- Provide appropriate signs.
- Provide a drinking fountain or cups.

Accommodation for clothing

- Accommodation for clothing shall be provided.
- Include a clothes drying facility.
- Provide separate facilities for men and women.

Facilities for rest

- Facilities for rest shall be provided.
- Separate smokers'/non-smokers' facilities to be provided.
- Separate facilities for pregnant women and nursing mothers to be provided.
- Provide a meal preparation and eating area.
- Provide a means for boiling water.

Regulation 23: Fresh air

Take suitable and sufficient steps to ensure the workplace and its approach has sufficient fresh or purified air. All fresh air plants are to have a failure warning device.

Regulation 24: Temperature and weather protection

Suitable and sufficient steps should be taken to ensure that during working hours the temperature at any indoor place of work to which these regulations apply is reasonable having regard for the purpose for which that place is used.

Every place of outdoor work is to be arranged so that it provides protection from adverse weather.

Regulation 25: Lighting

Provide suitable and sufficient lighting:

- at workplace;
- at approaches to workplace;

- on all traffic routes;
- where possible use natural light;
- consider colour rendition of artificial light;
- provide secondary lighting.

Regulation 26: Good order

All parts of any construction site which is used as a place of work to be kept in good order and in a reasonable state of cleanliness.

If necessary in the interests of health and safety, identify the perimeter of a construction site so that it is easily identifiable using suitable signs.

Prevent danger from projecting nails.

Regulation 27: Plant and equipment

All plant and equipment used in construction work shall:

- be safe;
- without risk to health;
- be of good construction;
- be of suitable and sound materials;
- be of sufficient strength and suitability;
- be fit for purpose;
- equipment shall be used safely;
- equipment shall be maintained properly;
- equipment is to remain safe.

Regulation 28: Training

Any person who carries out any activity involving construction work where training, technical knowledge or experience is necessary to reduce the risks of injury to any person, shall possess such training knowledge or experience or be under such degree of supervision by a person having such training knowledge or experience as may be appropriate having regard to the nature of the activity.

Regulation 29: Inspection

This regulation needs to be read in conjunction with Schedule 7 and collectively they create the following requirements.

Place of Work	Time of inspection
1 Any working platform or part thereof or any personal suspension equipment provided pursuant to paragraph 3 (or c) of Regulation 6.	1 (i) Before being taking into use for the first time and (ii) After any substantial addition, dismantling or other alteration and (iii) After any events likely to have affected its strength or stablility and (iv) At regular intervals not exceeding 7 days since the last inspection.
2 Any excavation which is supported pursuant to paragraphs (1), (2) or (3) of Regulation 12.	2 (i) Before any person carries out work at the start of every shift and (ii) After any event likely to have affected the strength or stability of the excavation or any part thereof and (iii) After any accidental fall of rock or earth or other material.
3 Cofferdams and caissons.	3 (i) Before any person carries out work at the start of every shift and (ii) After any event likely to have affected the strength or stability of the cofferdam or caisson or any part thereof.

Inspections are to be carried out by a competent person and must include plant equipment and materials.

Regulation 30: Reports

This regulation is to be read in conjunction with Schedule 8 and between them they create the following requirements. Particulars to be included in a report of inspection:

1 Name and address of the person on whose behalf the inspection is carried out.
2 Location of the place of work inspected.
3 Description of the place of work or part of that place inspected, including any plant equipment and materials.
4 Date and time of the inspection.

5 Details of any matter identified that could give rise to a risk to the health and safety of any person.

6 Details of any action taken as a result of any matter identified in paragraph 5 above.

7 Details of any further action considered necessary.

8 Name and position of the person making the report.

- Report to be contemporaneous.
- Copy of the report to be sent to the appointing person.
- Copy to be kept at the place of work for 3 months.
- Report to be made available to any inspector.
- Towers are exempt if erected less than 7 days.
- Twenty-four hour rule applies on scaffolding.
- Seven-day rule applies on excavation.

Regulations 31 to 35 are largely procedural, but for the sake of completeness their titles are listed below.

- Regulation 31: Exemption certificates
- Regulation 32: Extension outside Great Britain
- Regulation 33: Enforcement in respect of fire
- Regulation 34: Modifications
- Regulation 35: Revocations.

8.9 Further case studies

The new Construction Regulations remain reasonably simple to understand and use, to those who take the time and trouble to become familiar with them. It is important to remember that the regulations are not a standalone document. They are intended to be read in conjunction with HSE guidance and specifically, construction summary or construction information sheets. At any one time there are around 1500 of these in nine main sets, available free of charge by HSE.

Drawings and design information to a competent standard also play an important in achieving compliance with the Construction Regulations. Figure 8.2 illustrates the point.

It is not possible to dig a trench safely if the depth is unknown. Somebody has to take a decision about the depth of the trench. In most cases this should be the designer, who should specify the trench depth, but if this is not possible then the employer who is controlling the excavation must ensure that this is done. The point is illustrated by reference to the regulations, and the summarized version of them on previous pages.

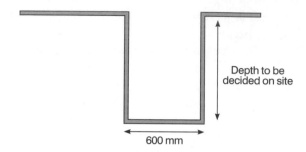

Figure 8.2 Case study: Construction Regulations explained

First, the trench is a place of work, so it must be kept safe (see Regulation 5).

Second, if the trench is more than 2 metres deep then Regulation 6: falls, might apply together with Schedules 1 and 2. If Regulation 6 applies, Regulation 8 almost certainly applies also.

If there are any structures near the trench – the design in Figure 8.2 does not deal with this point – then Regulation 9: Stability of structures, might apply.

Regardless of the trench depth, Regulation 12: Excavations, will apply and to some degree or other, all of the following could apply:

- Regulation 14: Prevention of drowning
- Regulation 15: Traffic routes
- Regulation 16: Doors and gates
- Regulation 17: Vehicles
- Regulation 19: Emergency routes and exits
- Regulation 20: Emergency procedures
- Regulation 22: Welfare facilities
- Regulation 23: Fresh air
- Regulation 24: Temporary and weather protection
- Regulation 26: Good order
- Regulation 27: Plant and equipment
- Regulation 28: Training
- Regulation 29: Inspection, and Schedule 7
- Regulation 30: Reports, and Schedule 8.

Whatever the depth, it would also be prudent to refer to HSE Construction Summary Sheet Safety in excavations, and to CITB Construction Site Safety Sheet 10, Excavations. Both give practical guidance on safe excavation, and refer to other relevant documents including British Standards (there are at least four) and CIRIA Report 97: Trenching practice 1983.

So, one apparently simple activity, digging a trench, could, depending on the depth, require compliance with 18 of the new Construction Regulations and four schedules, at least three other sets of regulations including CDM, the Management regulations and Lifting Operations, four British Standards, one set each of HSE and CITB guidance and one best practice manual in the form of the CIRIA Guide.

Figure 8.3 shows a typical arrangement that a factory owner might make, prior to leaving on annual holiday. At his 60 000 sq.ft manufacturing plant in the Midlands, he has arranged in his absence, as he does every year, for certain maintenance works to be carried out by his directly employed staff, and improvement work to be carried out by specialist subcontractors, all of whom are known to him. All work is to be started on 1 August and finished by 28 August.

At any one time, two of the factory owner's direct employees will be at work on the maintenance, and there could be up to 20 subcontractors employed on the improvement work. The factory owner's operations manager will be in overall control for part of the time, but is also away on holiday from 7th to 21st August. However, several meetings have been held with all concerned during July and the factory owner and the operations manager are entirely confident that everybody knows exactly what is expected of them and

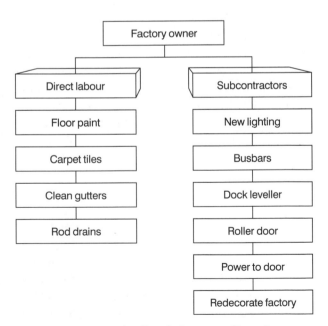

Figure 8.3 Case study: Construction Regulations, compliance issue

that they will simply get on with their jobs. There is some concern at the scale of this year's improvement works, which, at around £100000, are triple a typical year's expenditure.

On 12 August a subcontract painter falls 5 metres from a ladder in the factory and is seriously injured, requiring a subsequent 10 weeks off work.

What happens next? The factory owner would probably find himself in considerable difficulty under at least three sets of regulations.

First, under the CDM Regulations, the factory owner is the client. He has failed to appoint a planning supervisor and a principal contractor on what is clearly a CDM compliant project and could therefore be prosecuted for two breaches of Regulation 6. In addition he might have great difficulty in explaining how he had met the obligations of a client in respect of the following CDM Regulations:

- Regulation 8: Competence of designers and contractors
- Regulation 9: Adequacy of resources
- Regulation 10: Allowing work to start without a health and safety plan
- Regulation 11: Providing relevant information to designers.

Second, under the Construction Regulations the factory owner is an employer, and the person who controls the way in which the work is carried out. In the particular circumstances of the accident, both he and the painting subcontractor would be asked to explain the steps they took to comply, initially, with Construction Regulation 6: Falls.

It might be that the factory owner had agreed, either orally or in writing, that the painting subcontractor would provide and maintain his own access scaffolding. If so, this might provide him with a defence both to this charge, and to proceedings under Regulations 8 and 9 of CDM. On the other hand, if there was no such agreement, or if the factory owner had accepted responsibility for scaffolding then he might be liable for prosecution for a range of offences starting at Regulation 6 and embracing Regulations 5, 27, 28, 29 and 30.

Finally, compliance with the Management Regulations 1992 and in particular, Regulation 3: Risk assessment, might also be an issue if there were more than five persons involved in the painting operation at the time of the accident.

So in a worst case scenario, the factory owner could find himself facing eight charges under the CDM Regulations, six under the Construction Regulations and one under the Management Regulations, a total of 15 charges. The maximum fine in each case is £20000 so he might face fines totalling £300000 and he could also be jailed for up to two years.

He might then also be liable for the payment of damages to the painter under a civil action which might also run into tens if not hundreds of thousands of pounds.

8.10 Chapter summary and conclusions

The modernized Construction Regulations 1996 form a seamless interface with the CDM Regulations. Between them, responsibility for health and safety is placed upon those who design, manage and carry out construction work from project inception to project completion and handover.

The Construction Regulations are written in a very broad way and cover most, if not all activities which can take place on a construction site.

They are not a standalone instrument and are specifically intended to be explained by and read in conjunction with published HSE guidance. CITB Site Safety notes and CIRIA documents also provide detailed guidance on how to achieve compliance with the provisions of the Construction Regulations, and most other commonly occurring legislation affecting work on site.

9

Dealing with health and safety in construction contracts

9.1 Introduction and background

Many contractors are so used to receiving contract documentation containing long lists of terms and conditions that they have stopped reading them. Buried in these long lists there are often to be found clauses along the following lines:

> The contractor should be deemed to have included in his prices for any and all matters in connection with the following, and shall be solely responsible for all matters arising in the event of failure to comply...

A long list of the A–Z of Health and Safety legislation then follows.

Frequently, it is obvious that the clauses and the A–Z have been word processed from previous jobs, and within the A–Z many irrelevant or even repealed laws are written into the contract. By way of example, in 1998 the writer saw numerous examples of the Construction Regulations of 1961 and 1966 still referred to in building contracts awarded that year, despite the fact that they were repealed and replaced in September 1996 (see Chapter 8).

So it is pertinent to ask, what are the drafters of contract clauses of this nature trying to achieve? Why do they do it?

Two answers are usually given. First, they say they are trying to ensure that the contractor is aware of his responsibilities under health and safety law, and second, they are trying to give the employer a remedy in contract if the contractor fails to comply with any of his obligations.

On the face of it, these are two very sensible objectives, but are they necessary, and are they achieved by dealing with health and safety in contract in this way?

There is a school of thought that says it should not be necessary to spell out to a competent contractor his health and safety management obligations. Although there is a case for assisting him, particularly in a competitive tendering situation, to quickly get to the root of health and safety issues in a project, the contractor is not assisted by word processed lists and onerous contract clauses, and he is assisted by a properly prepared, project specific, pre-tender health and safety plan.

The **Derby City Council (1988)** case made clear that there is nothing in the HASWA 1974 that makes health and safety solely the responsibility of one party. Surely, therefore, clauses such as the one above are ineffective for two main reasons.

First, they are not a defence in any criminal proceedings. Second, to the extent their employers rely on them, they may give a false sense of security. The employer may believe he has successfully passed risk and responsibility to a contractor. The contractor, for whatever reasons, may do nothing about the risks which may eventuate through lack of action by both parties.

How then are the employer's best interests served by dealing with health and safety in contract in this way, and is there a better way?

9.2 Roles and responsibilities in the UK

Clear distinction is made between the criminal and the civil codes.

In the criminal code, responsibility rests where it falls, on the principle that the person who created the risk is responsible for it. Where more than one party causes risk, then more than one party may be held responsible. That was the lesson of the **Derby City Council** case, and has been reinforced many times since HASWA 1974 and Derby 1988.

Contractual devices cannot pass criminal code responsibility. Indeed, to the extent that clients use them and rely on them, arguably, they may be providing clear written evidence of the way in which the law may have been broken.

In 1998 the writer was asked to comment on the following. A national contractor received an invitation to tender for the design and construction of a new £65 million manufacturing plant and office building in the Midlands.

The head contract was to be JCT CD81, and the tender documentation comprised the Employer's Requirements, containing six or so 1:200 and 1:100 scale drawings, a general specification, information on direct contract packages and a schedule for pricing purposes. There was no pre-tender health and safety plan accompanying the documents, and the following clause appeared in the general specification:

> The main contractor shall be responsible for the preparation of the health and safety plan complying with the CDM Regulations which shall be submitted with the form of tender and the priced schedule. The main contractor shall ensure that the health and safety plan complies fully with all aspects of the CDM Regulations at all times and shall indemnify and save harmless the employer from any failure to comply with this obligation.

Several issues arise.

First, referring to Chapter 7, it is probably impossible for the main contractor to meet this obligation because he would simply not have the design risk information needed to produce a compliant Regulation 15(1–3) health and safety plan, and could not therefore respond with a compliant Regulation 15(4) health and safety plan. He might be able to produce something that looked like a health and safety plan but it could not be specific to the project and in any case, he is not the planning supervisor, whose duty it is to ensure that the Regulation 15(1–3) health and safety plan is prepared.

Second, if both parties agree to it, this clause is probably effective as a contractual device, in so far as it would give the employer a remedy in contract if the contractor fails to perform the obligation. But so what, if the contractor cannot physically perform the obligation?

Third, it would give the employer no defence against a criminal prosecution for his possible breach of CDM Regulation 10 which requires the client to ensure that a health and safety plan complying with Regulation 15(4) is in place before the construction phase of any project starts, in other words, before any construction work starts on site.

Fourth, if the client relies on the device and does nothing, then he would be in demonstrable breach of several of the CDM regulations and parts of the approved code of practice. Guidance Note 77 says that health and safety plans should be issued as part of the 'tender documentation or similar proposals'. In the case in hand, this was not done. It is the planning supervisor's job to ensure that the pre-tender health and safety plan is prepared, and this might be difficult here. It is the principal contractor's duty to produce a construction phase health and safety plan. At the time of tender, no principal contractor is appointed and it is to be remembered that a tendering main contractor is not a principal contractor unless and until appointed.

Fifth, health and safety plan compliance prior to the start of work on site is quite clearly a matter for clients, taking advice from the planning supervisor if required. The employer in this contract was also the client under the CDM Regulations. It was the writer's opinion that the clause in the contract demonstrated the client's intention to avoid his obligations under Clause 10 of the CDM Regulations by passing them to the main contractor.

It seemed to me that the employer/client and his advisors had got their legal codes confused. The words used in the contract to provide the employer with a remedy for breach of the civil code were the same words which, if relied upon, would demonstrate to the Health and Safety Executive the way in which the client may have broken the criminal code.

9.3 Consultants' contracts

In the criminal code the duties of consultants can be twofold. To the extent that they may be designers within the meaning of the CDM Regulations, then they must comply with Regulation 13 which requires them to apply the hierarchy of risk control to their design, co-operate with other designers and with the planning supervisor. If they have any site-based duties, then their obligation is to notice and do something about unsafe work practices on site. This does not extend to telling contractors what to do. For the most part, the duty will be discharged by drawing any apparently unsafe working practices to the attention of the principal contractor or main contractor as appropriate. CDM Guidance Note 68 refers.

In the civil code, head consultants are usually given three roles, the names of which can vary, but the essence of which is the same.

If an **architect** is appointed on full RIBA SFA 92 terms and conditions, the roles are:

- designer
- client's agent
- contract administrator.

If a consulting engineer is appointed on ACE terms and conditions the same three aspects, designing, client's agent and contract administration are given to the **engineer**.

In broad terms, the designer designs the works, the client's agent may vary the works (but not the contract) and the contract administrator, administers the terms of the contract and acts in a quasi-arbitral role in deciding, any disputes between the employer and the main contractor.

The 'architect' or 'engineer' is a powerful, pivotal role in both building and civil engineering contracts. The architect/engineer can act, within the limits of his authority as the client, and can instruct the contractor, who must comply. However, the same powers do not cross from the civil code to the criminal code, so the architect/engineer has no powers of instruction over the principal contractor in matters of health and safety and has no authority to act as the client under the CDM Regulations.

Sometimes, architects/engineers are appointed as client's agent under the CDM Regulations, but that is a separate appointment outside the scope of the contractual role which unfortunately has the same name.

Consultants not appointed to this role, for example quantity surveyors or structural engineers, have the established duty to notice and do something about unsafe work practices on site.

For appointments made under RIBA/SFA 95, the following obligations are placed on the architect:

- advise the client on the duties of a client under CDM;
- co-operate with the planning supervisor;
- pass information to the planning supervisor for inclusion in the plan;
- pass information to the planning supervisor for inclusion in the file.

If architects do not perform these obligations to a standard of reasonable professional skill and care then they may be in breach of contract, and the client has a remedy. However, SFA/95 is silent on the following further obligations of designers under CDM Regulation 13.

- apply the hierarchy of risk control;
- co-operate with other designers.

The RIBA 'Architect's Guide to Job Administration under the CDM Regulations 1994' is also silent on these points, yet they are of cardinal importance to the way the CDM Regulations are applied in practice, and in particular to the planning supervisor's role. The ACE and other consultant contract drafting bodies have adopted a similar approach. It is not clear why this is the case, but it leaves clients without a remedy if consultants employed on these standard forms fail to carry out these duties to a satisfactory standard.

Clearly this raises three difficult questions.

- What is a satisfactory standard of the application of the hierarchy of risk control?
- What is a satisfactory standard of co-operation between designers?
- Who decides what is satisfactory?

In order to avoid protracted debate on these issues, many clients ask consultants to warrant compliance with all of the duties of a designer, which then gives the client the possible entitlement of compensation, but not rescission in the event of breach. It still leaves the above questions unanswered, and it still leaves the planning supervisor with the job of agreeing with consultants:

- how the application of the hierarchy of risk control is to be demonstrated on an on-going basis;
- how designer co-operation is to be demonstrated on an on-going basis.

One way round this problem is to write appropriate procedures into the pre-tender (Regulation 15(1–3)) health and safety plan and then to make compliance with the plan a term of the warranty. In other words, do not rely solely on the published consultants' forms of appointment, because they do not afford clients a full measure of protection, nor do they facilitate good practice planning supervisor procedures.

9.4 Main contractors' contracts

The importance of the procurement route in configuring the health and safety management strategy was explained in Chapter 5. The role of principal contractor falls naturally to the main or managing contractor during the construction phase of most projects, but the role of planning supervisor during this phase is not quite so clear cut.

To illustrate the point, it is instructive to examine the approach taken by the Joint Contracts Tribunal, and to compare and contrast it with the approach of the Institution of Civil Engineers, when dealing with CDM duties in contract.

JCT's two main forms of contract are:

- **Standard Form of Building Contract 1980 Edition**
  This comes in six main versions and has a total of 28 separate options. It is in common use for the implementation of consultant-designed/

contractor-built buildings, the traditional route, complete with partial contractor-design options. Amendment 14, issued in March 1995, amends this form of contract in respect of the CDM Regulations and Practice Note 27 gives guidance on how to implement the contents.

■ **The Standard Form of Building Contract With Contractor's Design 1981 Edition (incorporating Amendments 1–7)**
This form is used for the implementation of contractor-designed and built buildings; Amendment 8 and Practice Note 27 deal with the implementation of the CDM Regulations.

Standard Form of Building Contract 1980 Edition

To set the ball rolling, JCT assume that in the traditional consultant-designed/contractor-built procurement route, the architect will be the planning supervisor, but make provision for it to be somebody else.

It is then stated that the term 'principal contractor' shall mean the contractor, but an alternative is contemplated.

It is made clear that references to the health and safety plan in connection with the principal contractor mean the Regulation 15(4) plan.

The employer undertakes to notify any changes in the planning supervisor appointment in writing to the contractor.

The employer undertakes to ensure that the planning supervisor will carry out the duties of a planning supervisor.

The employer undertakes to ensure that where the principal contractor is not the main contractor, then the appointed contractor will carry out all of the duties of a principal contractor.

Whilst he is the principal contractor, the contractor undertakes to carry out all of the duties of a principal contractor. In particular he undertakes to develop the Regulation 15(4) health and safety plan and to inform the employer of any changes to the plan.

The employer is given the task of notifying the architect and the planning supervisor of any changes to the plan notified to him by the principal contractor.

Any changes in principal contractor shall be at no cost to the employer, and shall not comprise grounds for an extension of time. (This contemplates that there may be two possible grounds for changing the appointment, mainly completion of a contract package or lack of competence.)

Upon receipt of a written request from the planning supervisor, the contractor undertakes to provide information for the health and safety file. He also undertakes to ensure that his subcontractors provide information for the file, subject to the planning supervisor's written request.

The contractor undertakes to work in accordance with the health and safety plan.

Provision of the health and safety file information to, and requested by, the planning supervisor is a condition precedent to practical completion.

Failure by the planning supervisor to carry out his duties is a relevant event and may give grounds for extension of time.

Where the principal contractor is another contractor, failure by the principal contractor to carry out his duties is a relevant event and may give grounds for an extension of time.

If the employer fails to ensure that compliance of the planning supervisor or the principal contractor (where he is not the contractor), this is a relevant event and may give grounds for an extension of time.

Failure to comply with the requirements of CDM is grounds for determination by the employer, and by the contractor. If the architect nominates any subcontractors, his nomination instruction must include a copy of the current health and safety plan.

Nominated subcontractors must provide the principal contractor with information required for the file, and completion of the subcontract is a condition precedent upon the fulfilment of this condition.

The Joint Contracts Tribunal's intention was to turn the duties of a principal contractor back-to-back into obligations placed on the main contractor, assuming that he is also appointed as principal contractor, and Amendment 14 certainly achieves this. It is to be noted that the duties of a principal contractor embrace CDM Regulations 15, 16, 17 and 18.

The law draws a distinction between a contract condition, which goes to the root of the contract and gives remedies including compensation and rescission, and a warranty, which whilst still important, does not go to the root of the contract, and therefore gives rise to compensation entitlement only.

Thus, compliance with the CDM Regulations by the employer and by the contractor are conditions, and failure to comply gives rise to grounds for determination. All other terms are warranties.

Two of the CDM Regulations also give an automatic right to civil action. They are:

- CDM Regulation 10 which requires clients to ensure that a compliant health and safety plan is in place before any work starts on site;
- CDM Regulation 16(2)(b), which requires principal contractors to ensure that only authorized persons are allowed access to the site.

In adopting this approach, JCT have taken into their contracts the same apparent anomalies that exist in the regulations.

The principal contractor or main contractor is not obliged to prepare the health and safety file, merely to provide information for inclusion in it. The regulations do not make anyone responsible for physically preparing the file; the planning supervisor is given the duty of ensuring that one is prepared, and the client must make it available once in his possession.

The health and safety file

In most cases, it makes sense to add the health and safety file preparation to the main contractor's obligations, as he is usually best placed to do it, and will almost certainly be contracted to prepare a 'building manual' or similar which usually contains most of the information that would normally be included in the file. However, if the contractor is asked to prepare the file, there are four golden rules.

1 A suitably worded amendment to the JCT head contract and Amendment 14 must be made.
2 The format and content of the file should be established by the client and the planning supervisor and clearly spelt out in any tender documentation or similar proposals, as though it were part of the permanent works, to give contractors the opportunity to price its preparation.
3 You get what you pay for. Clients should expect to pay a reasonable sum for a completed health and safety file, and would do well to have seen to have paid for a properly prepared and presented file to demonstrate their own compliance with Regulations 9 and 12.
4 The planning supervisor should ensure that there is a file to hand over to the principal contractor at the end of the pre-construction phase, and that appropriate procedures are in place for collecting information and developing the file during the construction phase. It is sensible to write these procedures into the pre-tender health and safety plan, with which the contractor warrants compliance. Even if the main contractor is not the principal contractor, it still makes sense to contract with him to prepare the file.

Health and safety plans

Many people write health and safety plans into the contract. JCT do not do this, and the RIBA in 'Architect's Guide to Job Administration' specifically advises against it. The case of *Yorkshire Water Authority* v. *Sir Alfred McAlpine & Son (Northern) Ltd (1985) 32BLR114* highlights the perils of incorporating contractors' method statements into contracts (the health and safety plan is arguably an extended method statement).

JCT require the contractor to warrant compliance with the health and safety plan, an approach which avoids most of the pitfalls of making it a contract document and still leaves the employer with a remedy if the contractor fails to comply.

If the plan is written into the contract, it places implicit obligations on the employer not to prevent the contractor from complying with his own method statements. In the Yorkshire Water case the employer had specified named supplier's goods which proved impossible to obtain because of insolvency. Thus the contractor was absolved from responsibility as it was physically impossible to carry out the works to the specification.

The contractor can also argue that such a situation could be resolved by the issue of an architect's or engineer's instruction under the relevant provision of the contract, and that failure to issue an instruction prevented the contractor from developing the plan, in itself grounds for claim. So it is not a good idea to write the plan into the contract, and warranting compliance with the plan renders the practice unnecessary.

JCT With Contractor's Design 1981

In dealing with design and build, JCT have offered two main alternatives.

First, they say that Amendment 8 substantially follows JCT 80 and Amendment 14. In other words, the preceding description of the traditional route. The architect is the planning supervisor, the main contractor is the principal contractor and all of the contract mechanisms in JCT 80 apply also to JCT WCD 81.

Second, JCT offers an alternative where the contractor is both the principal contractor and the planning supervisor for the duration of the construction phase. This creates three significant differences in approach.

- Whilst he is the planning supervisor, the contractor undertakes to comply with the duties of the planning supervisor (a warranty).
- Whilst he is the planning supervisor, the contractor undertakes to prepare and deliver to the employer the health and safety file (a warranty).
- Whilst the contractor is the planning supervisor he has the right of reasonable objection on health and safety grounds to any changes proposed by the employer.

This approach removes one area of uncertainty for clients, in so far as it creates single-point responsibility for the health and safety file. However, the contract is silent on the time for delivery of the file, and on a remedy for failure to comply, two seemingly inexplicable omissions given that the client might face criminal proceedings if he were to use a building without a file.

JCT/ICE compared and contrasted

The JCT approach may usefully be compared and contrasted with that of another important contract drafting body, the ICE, responsible for ICE 5th and 6th Conditions of Contract used extensively in civil engineering.

Whereas JCT have gone to considerable lengths to deal with the principal contractor's duties in back-to-back fashion through Amendments 8 and 14, and then to explain their reasons through Practice Note 27, the ICE have taken a very different approach.

They deal with CDM in one new clause, 72, applying through all of their published contracts, in broad terms as follows. It is assumed that the engineer is the planning supervisor and the main contractor is the principal contractor, and no other arrangement is contemplated. Both the planning supervisor and the principal contractor are then required to inform each other of any actions they are taking in compliance with CDM Regulations.

Any actions taken by the planning supervisor or the principal contractor are engineer's instructions pursuant to Clause 13 of the main contract. Alterations to the health and safety plan receive particular attention.

If the action to be taken arises from the contractor's own actions then there is no time or money entitlement.

If the action to be taken arises from some action outside the contractor's control, then there may be an entitlement to time and money but it is up to the contractor to demonstrate this in the usual way, through the mechanisms in the contract.

Provision is then made for a new Clause 73, special conditions, which may or may not be used in connection with CDM.

In the accompanying guidance note reference CCSJC/GN/March 1995, the sponsoring bodies ICE/ACE/FCEC, say that the new clauses do not attempt to interpret the regulations, but set out a framework to achieve compliance.

However, the practice note and the clauses themselves are silent on all of the issues dealt with in detail by the JCT, and in particular the thorny issue of who actually prepares the health and safety file.

The ICE's approach effectively means that the contractor and the engineer are left to battle it out amongst themselves in the usual way of adversarial contracting, where the contractor's opinion triggers the issuing of a notice to the engineer, who then acts as quasi arbitrator to decide the matter.

For example, the contractor, who is also the principal contractor, may believe that the planning supervisor, who is also the engineer, is delaying the progress of the works by failing to ensure that the designer, also the engineer, provides information for the health and safety file as and when needed, and that this was not reasonably foreseeable. The contractor must given written notice to the engineer of the planning supervisor's action, or in this case lack of it, and the designer's lack of action would also be relevant.

The contractor's notice is an engineer's instruction under Clause 13, and critics have argued that it is a nonsense to give contractors the power to issue their own engineer's instructions in this way.

Notwithstanding that, the engineer must now act fairly and impartially in deciding the contractor's notice in accordance with the contract. If any extra time or money is due, the engineer must award it. Critics argue that it is not possible for the engineer to discharge this obligation to a satisfactory standard of fairness and equity. Now, if the contractor disagrees with the engineer's decision he can, since the implementation of the Scheme for Construction Contracts (England and Wales) Regulations 1998 on 1 May 1998, request the appointment of an adjudicator and seek an adjudicator's decision.

Clearly the ICE drafting body is content with its approach as it remains unaltered. However, it is difficult to see how it works to the benefit of the client as it creates two areas of difficulty for them.

First, it internalizes CDM to the point where the contractor and the engineer are given yet another excuse for adversarialism and argument, to the exclusion of the client in what may be an area of paramount interest and importance to him and for which he may ultimately be expected to pay. The ICE appear to be leaving it to the court to provide any legal interpretation of CDM that might be needed, an attitude which pre-dates Latham's reforms of 1994 and now seems out of step with the times.

Second, it still leaves the client with the added expense of having to pay lawyers to deal with the contractual mechanisms needed to implement procedures within what is clearly a minimalist framework. In practice, if the client wants the contractor to prepare the health and safety file under an ICE contract, the contract must be amended to this effect.

9.5 Subcontractors' contracts

The standard forms of domestic subcontract for use with JCT head contracts are DOM 2 where the subcontractor has design responsibility and DOM 1 where he has none. The approach is the same. Subcontractors are given the responsibility of contractors under the CDM Regulations in back-to-back fashion so, in brief, they:

- undertake to comply with the obligations of a contractor under Regulation 19 (a condition);
- undertake to comply with the health and safety plan (a warranty);
- undertake to provide information for the health and safety file (a warranty).

Thus the main contractor has rights and remedies directly linked and proportional to the duty of contractors towards principal contractors, and the planning supervisor (see Chapter 7, p. 209).

Instructively, the drafters of the CDM Regulations anticipated the situation where the main contractor would be the principal contractor, and would be obliged to deal with other contractors on a project with whom he had no contractual relationship. This is quite commonplace in construction, arising mainly:

- where the client appoints direct trade contractors to undertake specialist work packages;
- where there are phased or sectional completion arrangements involving two or more main contractors on the main site.

In this scenario the principal contractor is given two statutory powers at Regulation 16, sufficient to enable him to achieve his own compliance through the actions of all appointed contractors whoever they are in contract with. There is a reciprocal mechanism at Regulation 19 which requires contractors to comply with the principal contractor's reasonable directions, and again, no contractual link is needed for these duties to be binding.

The 'blue form' is used on ICE contracts, and uses the same approach as the two main head contracts.

9.6 Contractor-designed works

JCT deal with the issue of contractor-designed works in a patchy way; ICE does not deal with it at all.

JCT 80 contemplates the traditional route, that is consultant-designed buildings, but in amongst its 28 options there are several that deal with partial contractor design.

In a nod in the direction of partial contractor design, Clause 42.2, performance specified works, is redrafted. Now a contractor's statement must be prepared and submitted to the planning supervisor, who is given the power to comment on the draft statement. The contractor must take account of the planning supervisor's comments before providing the architect with the final statement. The architect does not appear to have the right to comment upon the contractor's statement at any stage. The contractor must carry out the performance specified work in accordance with the statement. JCT explain that performance specified work contains design proposals and this mechanism is intended to ensure that the planning supervisor sees them before they are given effect, that is, before they become design.

This may deal with the issue of design compliance, but it is ineffectual in the matter of designer co-operation.

In the writer's experience, many contractors omit to submit the draft contractor's statement to the planning supervisor, and it is usually necessary to make a vigorous and sometimes prolonged intervention to obtain it. To reprise and paraphrase CDM Regulation 13, no work should be put in hand on site unless it has been design risk assessed, and demonstrated to the planning supervisor.

A somewhat more thorough approach is adopted to work designed by nominated subcontractors. In NSC/W Standard Form of Employer/Nominated Subcontractor Agreement, the nominated subcontractor, to the extent that he has designed his work, warrants that he will comply with CDM Regulation 13, Requirement's on Designer. Furthermore, the employer undertakes to ensure that the planning supervisor is provided with any nominated subcontractor design. Paradoxically, nomination of subcontractors is now a very rare event indeed. Most clients are advised against it.

In JCT CD 81, the contractor gives an undertaking that he has taken into account the requirements of CDM Regulation 13, Requirements on Designer, in any design he has prepared.

Under the CDM Regulations if the contractor prepares a design, or arranges for any person under his control to prepare a design – which is essentially what design and build contractors do – then he is unequivocally a designer and has a statutory duty to comply with Regulation 13.

It is a matter for conjecture why the JCT merely require contractors who are designers to 'take account of' Regulation 13. Quite what is meant by this is anybody's guess, and again it appears that the client is left without a remedy if the contractor/designer fails to comply either with CDM Regulation 13 or the shadowy requirement to take account of it.

It is baffling that nominated subcontractors, who are hardly ever used, must warrant compliance with Regulation 13, but squadrons of design and build contractors have the lesser obligation of 'taking into account', and legions of main contractors in the traditional route, who design CDM-compliant formwork, falsework and scaffolding and also design performance specified work, have no obligations at all in respect of CDM Regulation 13.

9.7 Suppliers

None of the main standard forms have been amended to take into account the position of suppliers.

Product, plant and equipment suppliers have general duties under HASWA 1974, and can have specific duties under various regulations including

COSHH, PUWER and the like. Compliance is a matter for them and is usually demonstrated by way of product data sheets, manuals, signs, etc. It is not usually necessary to deal with straightforward compliance with health and safety law in purchase orders or other contracts.

However, the situation can become less clear when the supplier is relied upon to provide design or design advice in relation to its product, plant or equipment.

CDM raises many questions about the role of suppliers in contributing to design. Is a kitchen unit manufacturer a designer within the meaning of the CDM Regulations? What about boiler suppliers, or window manufacturers, or roof truss suppliers? Are there standard details or product 'design' sheets, and if so, how should these be dealt with in contract?

It will probably take case law to bring a measure of clarity to these areas.

9.8 Benefits and pitfalls of contracts and health and safety law

If the main benefits of dealing with health and safety contracts are:

- to make clear to the parties their obligations and responsibilities;
- to provide a mechanism for ensuring compliance with health and safety law;
- in the event of breach, to give the injured parties a remedy to the extent that any of them are harmed by the breach,

then the main pitfalls must be:

- the contract does not make clear the obligations of the parties;
- the contract requires the parties to do less than is required of them by statute;
- the contract requires the parties to do more than is required of them by statute;
- the contract may be drafted in an unclear, ambiguous or onerous way that places responsibility with one party and authority with another.

The contract does not make clear the obligations of the parties

This can come about in a number of ways. One main way is when the obligations of the parties are not clear, or not fully known to them. Another way is where no written contract, or modified standard forms or bespoke contracts are used. Again the procurement route and contract strategy are of cardinal importance and the point is perhaps best illustrated by example. A case study is given in Chapter 8 where a factory owner undertakes work at his own factory using a combination of his own directly employed labour and specialist subcontractors. Many organizations make similar arrangements, and the point to note is that there are no consultants and no obvious main contractor in the matrix. It is used partly to save money by reducing the perceived unnecessary overhead (consultants and a main contractor) and partly to achieve greater customer satisfaction – the client knows exactly what he wants in terms of function and performance of the finished work.

Unfortunately, this arrangement does not quite fit with any of the recognized procurement routes described in Chapter 5. It has elements of at least two, namely, design and construction management, but fails the crucial test of single-point responsibility for both design and management. So the question arises, who is responsible for what?

Figure 9.1 reprises the Chapter 8 case study, adding some of the many 'invisible' but essential tasks that are part of the factory owner's undertaking.

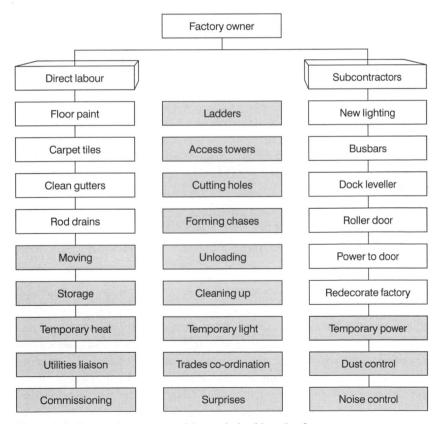

Figure 9.1 Case study: contractual issues in health and safety

In the factory owner's arrangements, operatives are usually left to get on with things as a matter of common sense. Perhaps some of the 'invisible issues' will be properly dealt with, but experienced contractors and operatives, most of whom have worked in similar arrangements at some time or other, will know that they will not, for two reasons.

First, no clear arrangements have been made and second, these invisible tasks cost money to undertake, and nobody is getting paid to do them. The real issue comes back to good management (see Chapter 4).

Somebody has to deal with the ordering, delivery, off-loading, erection, alteration, moving, dismantling and removal of ladders, scaffolding and access towers. In the case study these are essential to the tasks of cleaning gutters, new lighting, busbars and redecoration of the factory. The factory owner's direct labour and at least two subcontractors will use the access equipment but everybody will be affected by it in some way or another.

So the contract matrix is a good way of answering the question 'who is responsible for what?' by clearly spelling it out, but the same message has to be given to everybody. However, the health and safety plan is a better way of dealing with the issue, particularly in the absence of a contract matrix.

In the case study, it can be assumed that the factory owner made no such arrangements, partly because he does not know the full extent of what is needed (the invisible tasks) and partly because responsibility may not be clear. For example, who is responsible for disposing of the old factory lights? Also there may be no matrix of standard forms of contracts that can be applied to this arrangement.

The default position in the case study is that the factory owner has issued the subcontractors with his standard purchase order, complete with terms and conditions in small print on the back, along the following lines:

> Please carry out new lighting installation to the factory, all in accordance with your enquiry dated . . . . and your quotation dated . . . . for the lump sum fixed price of £ . . . . plus VAT. Terms net 30 days.

In this scenario, taking into account the essential invisible tasks, the full obligations of the parties are not made clear. Similarly the health and safety obligations are unclear. The default position is that the subcontractors are employers within the meaning of HASWA/MHASWA and the Construction Regulations, and are also designers and contractors under CDM. However no-one has appointed them as such. The fault rests firmly with the factory owner, and returning to the accident described in Chapter 8, he has no remedy in contract against the painting subcontractor for delay, loss and expanse arising from the accident.

The contract requires the parties to do less than is required of them by statute

Earlier in this chapter it was mentioned that in 1998, examples were seen where main contractors were obliged to comply with the now repealed 1961 and 1966 Construction Regulations. This requires them to do considerably less than is now asked of them by the new Construction Regulations which came into force in 1996.

An example may illustrate the point. Regulation 26, Good Order, is new to the 1996 regulations and requires that 'every part of a construction site shall be kept in good order. . .'. Instructively the regulation is silent on the matter of who actually holds the duty to keep good order, so it is to be assumed that everybody working on a site, or connected with work activity on it, may have a responsibility. In other words, the regulation has been drafted to give the Health and Safety Executive the opportunity to bring proceedings against anybody who they believe may have played a part in any accident arising where a site was not in good order.

In the factory owner case study, if an important customer visited site while the work was in hand and tripped and fell on a badly stored scaffold tube causing a notifiable injury, the factory owner could face criminal proceedings under this regulation, and could then face a civil action for damages from the injured customer.

Requiring compliance with repealed laws is the most common way of requiring parties to do less in contract than is required of them by statute.

The contract requires the parties to do more than is required of them by statute

It is often the case that contractors are required to submit method statements to contract administrators for comments or approval. There is no statutory obligation in any of the regulations applying to construction to prepare method statements. It is widely recognized as good practice, HSE strongly recommend the use of method statements, but the point remains valid. They are not required under statute but are frequently made an obligation under contract.

It is less common, but still happens, that contractors are required to submit written risk assessments for all activities they expect to undertake on a site. The statutory requirement is for risk assessments to be written where there are five or more persons involved in the work activity to which they relate.

Writing terms like these into contracts should be given careful considera-tion and should not be done unless there are very good reasons. For the most part, contract administrators do not have the training or the experience to make helpful comments on contractors' method statements or risk assessments, and has been pointed out elsewhere in this book, Guidance Note 68 of the CDM Regulations makes it clear that there is no duty on consultants or contract administrators to exercise a health and safety management function over contractors as they carry out work on site. The duty extends to noticing and doing something about unsafe work practices, no more and no less. It is unlikely that 'over specified' clauses such as these would impress an HSE inspector, or have any material effect on criminal proceedings.

The contract may be drafted in an unclear, ambiguous or onerous way that places responsibility with one party and authority with another

The issue of onerous contracts has been hotly debated in construction for decades. Main contractors complain about the contract terms imposed on them, often with considerable justification. However, they are less willing to debate the even more onerous terms and conditions that they, main contractors, impose on their subcontractors and suppliers as a matter of standard practice.

Typical sample subcontract clauses include the following:

> The subcontractor shall provide all of his own setting out, builders' work in connection with his trade and any attendances he may require.

In a small project for work of a simple nature, this clause may be efficient and effective. On a large or complex project it will be neither, because there will be a number of subcontractors all of whom have different requirements at different times in respect of setting out, builders' work and attendances in the form of temporary heat, light and power, scaffolding and access equipment, loading and off-loading and material distribution and so on. Making each subcontractor solely responsible for their own activities will result in duplication, waste and re-working on a large scale. All of these activities require planning, programming and co-ordination and management from a central source, the main or managing contractor. Giving the latter the power to make set-offs does nothing to address the operational requirements of those who must carry out the work on site. It also acts as a disincentive to the main/managing contractor's management and site staff who, instead of planning, programming and co-ordinating all of the subcontractors, usually find themselves in contention with individual subcontractors about who should have done what and when. The end result is usually inefficient and unsafe working.

> The subcontractor shall make his own co-ordination and co-operation arrangements with all of the other subcontractors on site and shall bear all possible consequences for failing to comply with this requirement.

Under the Management Regulations 1992, it is the responsibility of the main employer in a workplace to make adequate co-operation and co-ordination arrangements for health and safety with all other employers sharing the workplace. Under CDM Regulation 16, the duty is given to the principal contractor to perform a similar function with all of the contractors employed on the project whether or not he is in contract with them, in respect of health and safety arrangements. A clause such as this again acts as a disincentive to the principal contractor's management and supervisory staff, and to the extent that they rely on it and do nothing themselves, again arguably provides written evidence of the way in which the principal contractor might try to avoid his responsibilities by passing them to contractors.

Furthermore it is probably not possible for subcontractors to actually comply with the obligation placed upon them, as they may not know of each other's existence until they start work on site.

> In the event of the subcontractor's failure to comply with these obligations to the main contractor's reasonable satisfaction, he shall be entitled to have any unsatisfactory work taken down or done by others and shall be entitled to set-off any loss and expense incurred against any sums due to the subcontractor.

Whilst this clause gives the main contractor the authority to set-off loss and expense arising from breaches of the previous two, in turn the conditions are created whereby breach by the subcontractors is almost unavoidable. Thus it can be argued that, rather than acting as sensible risk transfer devices, these clauses actually increase the quantum of risk in the project to the detriment of all parties. The 'good order' clause, 26, in the Construction Regulations 1996 is a clear example of this. In the event of breach by one subcontractor, everybody else working on the site could be in the frame of any Health and Safety Executive investigation, and the main contractor may have some awkward questions to answer about his own actions in the matter.

Further pitfalls of dealing with health and safety in contract include:

- writing health and safety plans into contracts;
- dealing with changes;
- planning supervisor interventions;
- what happens in the event of breach?

Writing health and safety plans into contracts

The ultimate purpose of health and safety plans is to provide information to those working on construction sites about risks to their health and safety. They are not meant to be contract documents and the **Yorkshire Water** case highlights the 'double-edged sword' nature of the mechanism brought into existence when this is done. The JCT approach of requiring contractors to warrant compliance with the health and safety plan appears to offer all parties an effective remedy in the event of non-compliance.

Dealing with changes

Changes in the works are a fact of life in construction. All of the standard forms envisage that the client/employer will change his mind, or that it may be found necessary to make changes to the design whilst construction work is proceeding on site. The standard forms all deal with this, applying the basic principles that the works may be varied, but not the contract, and the contractor and his subcontractors are entitled to be paid what the changed work is worth in terms of time and money (as distinct from what it actually costs them). It is to be remembered that CDM Regulation 13, Requirements on Designer, stands alone, that is, it applies to all construction design all of the time. If it is found necessary to vary the design of the works, then before any change or variation order is issued through the mechanisms of the contract, great care should be taken to ensure that the design and designers are fully compliant with Regulation 13, in other words that designers have co-operated and applied the hierarchy of risk control to their design. The RIBA recommend that any architect's instructions issued by contract administrators are endorsed to this effect, and that any new design risks are set out on the instruction. This is a useful device and informs both the planning supervisor, who may have to alter the Regulation 15(1–3) health and safety plan, and the principal contractor who may have to respond with a revised Regulation 15(4) health and safety plan.

Planning supervisor interventions

Unlike the principal contractor, the planning supervisor has no powers under statute to issue instructions. He is not named as an officer of any of the standard forms of contract, so he has no powers of instruction there either, yet as Guidance Note 80 of CDM makes clear 'in some circumstances, the planning supervisor may need to be proactive and decide on the best options for health and safety within the constraints of the projects'.

How then is the planning supervisor to enforce his reasonable requirements if he finds he needs to be proactive and decide an option? Should he be given powers in the contract? The answer must be 'no'. Planning supervisors must persuade contract officers to issue any instructions that might be needed to implement their reasonable requirements.

Often, the contract officer is also the planning supervisor, but he must make a clear distinction between the roles and be seen to keep them separate.

What happens in the event of breach?

The short answer is 'Nobody really knows'. The general principle of 'no contracting out' is well established under the **Unfair Contracts Terms Act 1977**, but this relates to the relatively narrow field of liability for negligence causing personal injury.

The **Derby City Council** case still sets a benchmark for health and safety in construction contracts, where the judge simply looked straight past the words in the contract and decided the case on the basis of the facts and the law, stating that there was nothing in HASWA that made health and safety the responsibility of one party. However, that case and principal was decided before the Management Regulations, CDM and the new Construction Regulations came into force, and it may be that further test cases involving these regulations may modify the law.

However, it is difficult to see how the principle that health and safety is the responsibility of all involved parties can be altered. It follows, therefore, that clauses in contracts which seem to place responsibility solely with one party must be ineffective, and if they are, then the remedies – set-off, etc. – cannot be applied. In order to apply them, it must be necessary to demonstrate breach of contract, which in turn is demonstrated by breach of statute, which only a criminal court can decide. In other words, if there is no criminal conviction there can be no breach of contract.

Therefore there seems little point in writing such clauses into contracts. To the extent that health and safety is dealt with in detail in construction contracts, the principle must be that statutory duties must be back-to-back with contractual obligations, no more and no less.

9.9 Chapter summary and conclusions

Dealing with health and safety in contract is not straightforward. The **Derby City Council** case established that contract terms and conditions provide no defence in a criminal action. The lessons of Derby remain largely misunderstood and contract drafters continue to try to achieve single point responsibility by transferring it from clients to main contractors, who in turn, often seek to transfer their health and safety responsibilities to subcontractors.

The introduction of the CDM Regulations has added to the complexity of the matter and has been dealt with in an inconsistent way by the main contract drafting bodies.

Consultants' contracts are silent on designers' obligations under CDM Regulation 13.

Main contractors' contracts issued by JCT adopt a back-to-back approach in turning statutory duties into contractual obligations, but they do so in an incomplete way and they leave the issue of the health and safety file preparation unaddressed, unless a design-and-build contract is used.

JCT subcontracts vary considerably in their treatment of subcontractor-designed work.

ICE main contracts and subcontracts deal with CDM in one short clause which does not seek to interpret the law.

There are three main benefits claimed from dealing in detail in contract with health and safety (see Section 9.8).

There is anecdotal and written evidence that contract drafters often make mistakes in the amendments they make on health and safety grounds. When this happens, there can be at least three, and sometimes more major pitfalls. To the extent that health and safety is dealt with in contract, it is sensible to apply the back-to-back principle in relation to statutory duties and contractual obligations.

10

The cost and price of health and safety

10.1 Introduction

In 1860, John Ruskin wrote:

> It's unwise to pay too much, but it is worse to pay too little. When you pay too much you lose a little money – that is all. When you pay too little, you sometimes lose everything, because the thing you bought was incapable of doing the thing it was bought to do.

Guidance Note 57 of the CDM Regulations says:

> The cost is counted not just in financial terms but also in those of fitness for purpose, aesthetics, buildability or environmental impact.

The cost referred to is the cost of applying the hierarchy of risk control to the designing of construction, but the broader debate must embrace the cost of managing health and safety within the overall construction process (which includes designing), and how it is calculated and expended.

10.2 Pricing health and safety construction

At some stage, someone has to put prices to the measures needed to manage health and safety in construction projects. The question is, how is this to be done?

Chapter 5 examined a number of paradigms which configure the construction process. It may be instructive to revisit them, namely:

- lowest price mentality of clients;
- competitive tendering;
- dutch auctioning;
- adversarial contracts;
- subcontracting;
- design separation.

This chapter will now examine the way in which the above impacts on health and safety management arrangements in construction projects, and seeks to answer several important questions, namely:

- How should construction be priced?
- How is it actually priced?
- How does the pricing process affect safety in design and on site?

10.3 How should construction be priced?

The basic premise upon which all work, including construction, should be priced is, first to ascertain the work to be done, second to calculate the actual cost of carrying out the work and third, to add on a reasonable margin for general overheads and profit.

Once again, the importance of the project management structure and the procurement route and contract strategy comes to the fore. Previous chapters have examined the relationships between project management structures, choice of procurement route and the contract strategies needed to implement both. However, the starting point in pricing construction is ascertaining the work to be done, and this falls into two broad categories.

- designing;
- constructing;

Designing by consultants

The amount of designing work in a project is initially estimated by reference to the client's design brief and budget.

Designing is a labour intensive activity which usually requires suitable premises and benefits from the use of appropriate tools and techniques such as computer-aided design. Consultants provide design and related services such as cost advice, and until about 1986 it was standard practice to employ consultants on the scale fee recommended by the relevant professional institution. In that year the Director General of the Office of Fair Trading began a campaign to discourage the use of scale fees, and since then, thanks to compulsory competitive tendering in the public sector and a much more competitive approach from consultants from the private sector, scale fees are a thing of the past. Now consultants are expected to compete for work, and as everybody knows, notwithstanding Ruskin's words of wisdom, once into the bidding arena, only price – the lowest price – matters.

So how should health and safety management arrangements be priced by consultants in design appointments? Guidance can be found at Regulation 9, Provision for Health and Safety, of the CDM Regulations, and in particular Guidance Note 36 which says:

> In checking as to the allocation of resources, . . . resources is a general term which includes the necessary plant, machinery, technical facilities, trained personnel and time to fulfil the obligations.

Further guidance, which is also relevant to all other consultants' agreements, can be found in the Engineering and Construction Contract (ECC), specifically the guidance notes accompanying the Professional Services Contract.

Helpfully the PSC guidance explains what is meant by 'actual cost' as follows:

- Staff rates which include:
  - basic salaries;
  - additional payments or benefits;
  - social costs including insurances and pensions;
  - office expenses including rent and rates;
  - administrative staff.
- time expended on services;
- expenses.

The PSC guidance contemplates a simple calculation as follows:

- estimated time to be expended on services,
- multiplied by staff rates,
- plus estimated expenses.

Thus the estimated net cost of carrying out the work is ascertained. The PSC then contemplates that a **fee percentage** will be added to the net cost.

This is added both to staff costs and expenses, and covers general overheads (those not contained in the staff rates) and profit.

PSC guidance explains that staff rates can be established in three ways:

- rates for named staff;
- rates for categories of staff;
- rates related to salaries paid to staff.

The PSC itself offers four main options for payment which are:

- priced contract with activity schedule;
- time-based contract;
- target contract;
- term contract.

Again these payment methods are applicable across the spectrum of consultants' appointments. The point made here is that the PSC offers a useful source of best-practice guidance for pricing consultants appointments in all commonly occurring situations, and is applicable not just to the Engineering Contract, but to all other building, civil engineering and process engineering contract matrices.

Most consultants calculate standard staff rates along the lines given in Table 10.1. The net staff rate must therefore include all of the necessary plant, machinery, technical facilities and trained personnel mentioned in CDM Regulation 9, Guidance Note 36. Calculations for time to be expended on the service must include estimated staff hours in accordance with the advice given at Guidance Note 36.

Table 10.1 Calculation of standard staff rates

Staff grade	Net staff rate (£)	Expenses (£)	Fee percentage	Total (£)
Partner or director	40.00	5.00	100	90.00
Associate	30.00	5.00	100	70.00
Senior consultant	25.00	5.00	100	60.00
Consultant	20.00	5.00	100	50.00
Technician	15.00	5.00	100	40.00
Administrator	10.00	5.00	100	30.00

Essentially, consultants should be able to demonstrate, if requested, that they have priced the duties of a designer as described at CDM Regulation 13, namely:

- inform clients of their duties, if applicable;
- apply the hierarchy of risk control to all design;
- co-operate with other designers;
- co-operate with the planning supervisor;
- provide information for the health and safety file.

There is no explicit statutory requirement for consultants to price these duties separately. Indeed some consultants have contended that it is not possible to do so, whilst others take the opposite view.

In my opinion, consultants can and should allocate resources against each CDM duty, and price them accordingly for three reasons.

First, it is good practice which provides a benchmark against which to measure performance on the project in hand, and in the future.

Second, the duties do actually take time and resource which may or may not be an additional cost burden on the job. That is not the point.

Third, in the event of an HSE investigation, separate pricing may be accepted as conclusive evidence of attempted compliance. The duty is to comply with the requirements on designers by allocating adequate resources. The duty is not to allocate **correct** resources, for who, other than a designer, is to say what is correct in terms of project-specific appointments. Accordingly, in my view, any criminal investigation can be expected not to examine correctness, but adequacy. If there is nothing in writing, consultants may find it difficult to demonstrate how they allocated adequate resources, and the persons appointing them may find it difficult to explain how they had checked on the allocation of adequate resources.

Designing by contractors

All of the same principles explained above apply equally to contractors who prepare CDM-compliant design. They should be seen to allocate adequate resources to their duties, and to price them separately. It is a relatively simple task.

Fitness for purpose, aesthetics, buildability and environmental impact

These are issues to be taken into account, and also into pricing, by designers when preparing designs. Sustainability is a term often used to describe all of these concepts, 'green' architecture is another, and there are said to be six principles of green design, which are:

- conservation of energy;
- work with climate;
- minimize new resources, re-use existing;
- respect for users;
- respect for the site;
- holism.

In other words, designers are expected to balance health and safety considerations against the wider requirements of good, sustainable design. For example, in restoring an old listed building, a designer would not be in breach of CDM Regulation 13 if he or she specified lead-based paint, or sand-lime mortar if it had been previously used on the restored building. However, it might be more difficult, but certainly not impossible, to justify such use on a new building.

Constructing

Constructing is usually carried out by contractors, including main contractors using a combination of directly employed and subcontracted labour, and construction managers and management contractors who use only subcontractors.

In order to ascertain the amount of work to be done on site, reference must be made to the design information which should allow contractors to clearly understand the nature, complexity, scale, scope and quantity of work and the quality standards to be achieved. Upon receipt of this information the contractor must consider ways of pricing it. However, the procurement route will configure the time at which contractor involvement is achieved and in the management routes the contractor may indeed be responsible for providing this information himself. Relevant guidance on pricing includes the following:

- CDM Regulation 9, Guidance Note 36 – describing adequate resources;
- Engineering and Construction Contract Guidance Notes, which again can be applied to all project situations;
- The Chartered Institute of Building Code of Estimating Practice.

The ECC Guidance Notes describe the word **price** in the following way:

- the contractor's estimate of actual cost plus other costs, overheads and profit to be covered by his fee;
- the contractor tenders his fee in terms of a fee percentage to be applied to actual cost.

Six types of payment mechanism are identified, namely:

1 priced contract with activity schedule;
2 priced contract with bill of quantities;
3 target contract with activity schedule;
4 target contract with bill of quantities;
5 cost-reimbursable contract;
6 management contract.

These do not belong exclusively to the engineering and construction contracts, they are simply names given to methods which have been around for a long time.

In a priced contract, the contractor is paid for work at the prices he sets out in his tender, whether they are fixed price sums (activity schedules) or unit rates (bills of quantities).

In a cost-reimbursable contract, the contractor is paid his 'properly expended costs'. This is not to be confused with a 'cost plus' arrangement.

Options 1–5 cover all of the commonly used methods of payment, not just under the Engineering Contract, but across the broad spectrum of construction under other standard forms and most workable bespoke contracts.

Outline methodology

The Chartered Institute of Building Code of Estimating Practice suggests that the main way of calculating prices for construction work is by the use of all-in rates for labour and plant, and unit rates for materials. The steps to be taken are:

■ Calculate all-in labour rate.
■ Ascertain amounts of work to be done.
■ Apply production standards or outputs to ascertain demands for work.
■ Multiply the amount of work to be done by the unit rate to obtain a total.
■ Repeat the exercise for plant.
■ Calculate or obtain material supply prices and convert purchase units to ascertain quantities of work to be done (for example, how many bricks per square metre) and allow for waste.
■ Multiply the ascertained amount of materials by the supply prices to obtain a total.

The next step is therefore to examine the above in more detail.

Calculating all-in rates for labour

The CIOB Code of Estimating Practice recommends the use of all-in rates for labour. The concept is widely recognized in construction, but the CoEP gives useful guidance on what should be included in a unit rate as follows:

- guaranteed minimum wages;
- bonus;
- inclement weather time allowance;
- non-productive overtime costs;
- sick pay;
- trade supervision;
- Working Rule Agreement allowances;
- CITB contributions;
- National Insurance contributions;
- holiday credits;
- tool money;
- statutory severance pay;
- employer's liability insurance.

The Working Rule Agreement currently states that a 39 hour week should be worked, comprising Monday to Thursday 8 hours plus 7 hours on Friday. Public and statutory holidays amount to 21 days and adjusting for weekends, this means that the normal hours available for work each year amount to 1933 per operative employed.

Taking all of the above into account and allowing one supervisor to two operatives, the CIOB CoEP calculated that at rates and prices prevailing in 1996/7 the cost to employers of directly employed labour was:

- **Craftsman**: £7.14 per hour (£13 082.89 per annum)
- **Labourer**: £5.57 per hour (£10 213.92 per annum)

A similar methodology is applied to calculating all-in rates for plant, and material prices are usually obtained directly from suppliers.

Production outputs, all-in rates and unit rates

The next step is the very nub of estimating. The contractor must now decide the outputs in respect of each separate item of work in the project which he believes he will achieve, and he must do this under the three headings of labour, plant and materials.

For example, there may be an item of work as follows:

Concrete, Class C35, in trench not exceeding 2 metres deep.

The contractor may choose between mixing the concrete on site, or purchasing it ready mixed from a supplier. Assume he chooses the latter, then he may estimate the outputs given in Table 10.2.

Thus the price of £71.14 represents the contractor's estimate of the net cost to him of buying and placing $1\,m^3$ of Class C35 ready mixed concrete in a trench exclusive of site based and head office overheads and profit. The price of £71.14 is described as the unit rate.

Unit rates and quantities

The contractor then takes the quantities of all of the work to be done on site, obtained from the design, and extends them at his unit rates. A prudent contractor will undertake this exercise by work activity or trade, that is,

Table 10.2

Work activity	Output	All-in rate (£)	Unit rate (£)
Discharge from truck and place concrete in trench: labourer	2 hours	5.57	11.14
Part time assistance from JCB and driver: plant	0.5 hours	20.00	10.00
Purchase price of concrete plus 5% allowance for waste	$1\,m^3$	50.00	50.00
		Net unit rate	£71.14

Table 10.3

Code	Unit	Description	Unit	Qty	Rate (£)	Amount (£)	Remarks
C10	1/101	Concrete Class C35 in trench not exceeding 2 m deep	m^3	100	71.14	7114.00	
	1/102	Concrete Class C35 in trench not exceeding 3 m deep	m^3	50	80.00	4000.00	

earthworks, concrete, formwork, brickwork and so on. This enables him to set and monitor budgets against trades more easily and also to begin to assemble a price for the whole of the project.

Analytical estimating

If the contractor stops here, he will know the quantities of work to be done and the total prices he expects to pay to have it done, but he will not know the composition of those prices, the labour, plant and materials contained within them.

In other words, he will not know or be able to demonstrate the adequate resources he will allocate to carrying out the work, including those needed to address his health and safety obligations. Nor will he be able to ascertain the time and money budgets from which the project planning standards flow, and which are needed to enable him to control the works. So one further step is needed as illustrated in Table 10.4. The contractor must break down or analyse his prices as shown in the Table.

In this way, the contractor can build up a complete resource picture of the project which he can apply to any or all of the payment options set out earlier in this chapter. He can also use the information for a number of different purposes including budget setting, planning, programming and monitoring and the ordering of materials.

Building up the tender total

From here, the Code of Estimating Practice recommends that tender prices are built up as shown in the box on p. 281 opposite.

Table 10.4

Code	Unit	Description	Unit	Qty	Rate (£)	Labour (£)	Plant (£)	Materials (£)	Net total (£)
C10	1/101	Concrete Class C35 in trench not exceeding 2 m deep	m^3	100	71.14	1114.00	1000.00	5000.00	7114.00
C10	1/102	Concrete Class C35 in trench not exceeding 3 m deep	m^3	50	80.00	1000.00	500.00	2500.00	4000.00

Net cost of estimated labour

Net cost of estimated plant

Net cost of estimated materials

Net cost of estimated site overheads

Net cost of subcontracted work

Total net costs   £_____

Scope (plus or minus)[1]

Risk

Head office overheads

Profit

Subtotal   £_____

Provisional sums[2]

Dayworks[3]

Tender total   £_____

[1] Scope indicates the contractor's willingness to deduct possible savings or discounts from his bid in order be more competitive
[2] Provisional sums are used where design is still at concept stage and a best wild guess is made of what the out turn cost will be. The mechanism is intended to ensure consistency and comparability in the assessment of competitive tenders.
[3] Dayworks can be thought as a contingency for unforeseen circumstances

The important points to note are these:

1 The methodology suggested by the Code of Estimating Practice can be applied to all construction work carried out by all types of contractor and subcontractors, and is applicable to all pricing and payment options.
2 The methodology again contemplates that contractors will estimate the net cost to themselves of carrying out the works, that is the prices to be paid for the necessary labour, plant and materials.
3 Contractors are then expected to make appropriate additions initially calculated as lump sum prices, but usually expressed as percentages of the total net costs in respect of scope, risk, head office overheads and profits.

The methodology should therefore deal with all of the 'adequate resources' mentioned in CDM Regulation 9, Guidance Note 36 which contractors have allocated to dealing with their health and safety management obligations (Figure 10.1).

Therefore a valid question is: does this methodology demonstrate adequate resources to the standard envisaged by the CDM Regulations?

A cautious answer is that only time and a few test cases will tell. However, in the writer's opinion, the answer to the question is 'no'. The methodology does not demonstrate adequate resources to the standard envisaged by the CDM Regulations. There is at least one, vital, missing ingredient in the form of the health and safety plan.

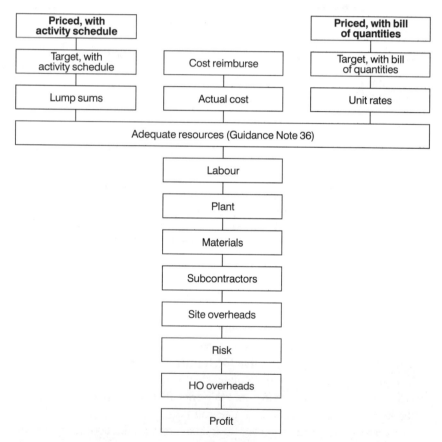

Figure 10.1 Pricing model. The model incorporates ECC Guidance, CDM Guidance Note 36 and the CIOB Code of Estimating Practice

CDM Regulation 15, Guidance Notes 77 and 78 make it clear that a pre-tender health and safety plan must be included in any tender documentation or similar proposal, for example, a works order or a negotiated package. During the tendering or pricing process, prospective principal contractors are expected to place their tenders upon the health and safety plan prepared by the planning supervisor.

Furthermore, Guidance Note 84 makes it clear that the principal contractor must develop the plan to incorporate seven main requirements, namely:

■ approach to health and safety management;
■ risk assessment under Regulation 3 of MHASWA;
■ common arrangements of client or principal contractor;

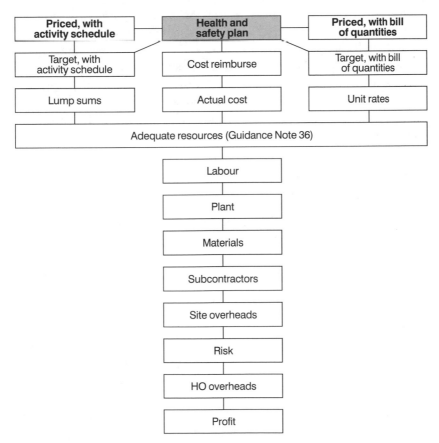

Figure 10.2 Pricing model with health and safety plan. The model incorporates ECC Guidance, CDM Guidance Note 36 and the CIOB Code of Estimating Pratice

- arrangements for filling principal contractor's duties;
- health and safety compliance monitoring arrangements;
- health and safety management roles;
- modification arrangements as work proceeds.

So it is the whole plan, embracing Regulations 15(1–3) and (4) which must be taken into account when pricing construction work on site.

Clearly, in order to achieve this outcome, prospective principal contractors must develop the Regulation 15(4) plan to the fullest extent possible, and take it into account during the tendering or pricing process. Figure 10.2 illustrates the point. Anything less than this may place principal contractors in a difficult position if they are called upon to demonstrate how they have allocated adequate resources in line with CDM Regulation 9. In turn, this may impact upon clients under Regulation 10.

One important issue flowing out of this is that prospective principal contractors must carry out risk assessments during the tendering or pricing process and before they finalize their prices. If they do not, then it may not be possible to calculate the adequate resources upon which to base net tender prices and the accompanying mark-ups for overheads and profits.

It is submitted that this is a further illustration of the reason why it is not possible to treat health and safety as some separate issue to the mainstream construction process.

Commercial attitudes

Many hard-pressed estimators, commercial managers and company directors might disagree strongly with some of the points made above. A number of opposing views can be put forward, namely:

- **Time**: there is insufficient time to carry out risk assessments during the tendering process.
- **Cost**: it would add unnecessary cost to the already expensive tender process.
- **Uncertainty**: the average success rate in price-based tender competitions is between one in six and one in ten. For that reason, contractors do not always calculate tenders to the Code of Estimating Practice standard because they know they have only a small chance of winning the job.
- **Unnecessary**: there is nothing within the Management Regulations or CDM that says that risk assessment should be carried out at the time of tender.
- **Unsuited**: it is sometimes said that estimators are not the right people to carry out risk assessment, due to lack of training.

I do not agree with any of these views. The time and cost argument can be set aside by reference to the Code of Estimating Practice, under the 'labour element' section, which lists 18 checkpoints to be used in calculating all-in hourly rates. If all 18 are dealt with, this would constitute a suitable and sufficient risk assessment under Management Regulation 3 in most cases. The only extra work estimators would need to do is to write it down.

The uncertainty point is, in my view, no reason to cut corners in the preparation of tenders.

The unnecessary issue is, in my view, mistaken. It has already been demonstrated that a condition precedent to the production of a compliant Regulation 15(4) health and safety plan is the inclusion of risk assessments in the plan. The real difficulty is prosaic. It is unlikely that subcontractors and suppliers could be persuaded to produce risk assessments at the tender stage.

The unsuited point may well be right, but that does not mean that estimators should be allowed to remain unsuited. There is plenty of good training available for them.

A further strong opposing view is that the site manager is the person best placed to carry out risk assessments, and that this process should be done near the time the work is to be started on site. This point can be easily accommodated in two main ways.

First by involving the site manager in the preparation of the estimate or tender, which has commercial and operational benefits, and second, by having the manager re-visit all risk assessments before starting any new activity on site, which is already a requirement of the Management Regulations anyway.

What more can be done during tendering?

The purpose of risk assessment is to eliminate, reduce or control risks. Therefore the allocation of adequate resources of identified control measures is cardinal to the risk assessment process.

Contractors should price the allocated resources into the all-in and unit rates that are used in the analytical estimate and preparation of the net tender sum.

It may also be appropriate to allocate priced items into the site or head office overheads. Many contractors already do this to a greater or lesser extent. However, many do not or cannot because they employ mainly subcontractors and rely on them to provide lump sum prices without breakdowns, which the main contractor simply incorporates into his own bid. When this happens the main contractor cannot know whether or not adequate resources have been allocated at the time of tender.

It is also always argued that adequate provision is made in the preliminaries section of bills of quantities prepared under recognized standard methods of measurement. From the point of view of the person preparing the preliminaries, this may well be the case, but the issue here is not how bills of quantities are prepared, it is how the items contained within the bill are actually priced. The use of 'preliminaries' as a means of pricing health and safety is recognized custom and practice, but it requires the contractor to treat health and safety as some separate issue, and as has repeatedly been made clear in this book, this is not an approach favoured by the writer.

Possible antidotes

The issue can be addressed at source. Using the risk assessment methodology set out at Chapter 6, the estimating team, which should include the prospective site manager, should identify the adequate resources needed to implement any specified risk control measures. These should then be priced into the developing tender, as illustrated in Figure 10.3.

The unit rate is the same as the previous example but the contractor could now demonstrate adequate resources against a compliant risk assessment. He also has the information to hand to prepare a compliant Regulation 15(4) health and safety plan. Rather than adding time and money to the tender process, this methodology demonstrably saves both.

Pricing note: Code C10: Unit 1/101		Job no:		
Concrete Class C35 in trench not exceeding 2m deep	Output	Rate £	Amount £	
Discharge from truck and place concrete in trench				
Basic labour	1.50			
Erect safety barrier/dismantle	0.50			
	2.00	5.75	11.14	
Part time assistance from JCB and driver				
Basic plant	0.25			
Keep site in good order	0.25			
	0.50	20.00	10.00	
Concrete price at cost		46.00		
Waste 5%		2.30		
Safety barrier hire cost		1.70	50.00	
		Net unit rate	£71.14	

Figure 10.3

Table 10.5

Code	Unit	Description	Unit	Qty	Rate (£)	Labour (£)	Plant (£)	Materials (£)	Net total (£)
C10	1/101	Concrete Class C35 in trench not exceeding 2 m deep	m^3	100	71.14	1114.00	1000.00	5000.00	7114.00
		(Safety				371.3	500.00	170.00	941.30)

In the analytical estimate the safety measures could appear as shown in Table 10.5. This could be reflected throughout the remaining stages of the tender price build-up, so that the adequate resources allocated by the contractor and his subcontractors are both particularized and summarized at each stage, ready for inclusion in the construction stage plan.

Quality price mechanisms

In recent years, post Latham (1994) and Egan (1998), there has been a marked shift away from lowest-price competitive tendering.

Most recognized authorities recommend that competition should be based on value for money, using quality price mechanisms as the basis for assessing value. Important guidance is offered in the following publications:

CIRIA: Special Publication 117 (1994): Value by Competition; A Guide to the Competitive Procurement of Consultancy Services for Construction.

Construction Industry Board Report No. 3: Code of Practice for the Selection of Main Contractors and Subcontractors.

Construction Industry Board Report No. 4: Selecting Consultants for the Team Balancing Quality and Price.

HM Treasury, Procurement Group, Procurement Guidance No. 1: Essential Requirements for Construction Procurement.

HM Treasury, Procurement Group, Procurement Guidance No. 2: Value for Money in Construction Procurement.

HM Treasury, Procurement Group, Procurement Guidance No. 3: Appointment of Consultants and Contractors.

10.4 How is construction priced?

Estimating and tendering methods are a jealously guarded secretive area of contracting across all sectors. It is difficult to know whether or not all of the larger contractors, for example, use similar estimating systems and methods, implemented to high standards. One imagines that they do, but experience continually demonstrates the opposite. In 1998, no less a contractor than Laing sheepishly confessed to a £20 million plus mistake in the tendering process for the new National Stadium for Wales in Cardiff.

Smaller contractors and specialists have the same range of software and published guidance available to them and again, the impression is that they should be on the ball, because of the more specialized and focused nature of their businesses. Again reality intrudes, harshly at times.

Pricing construction has always been a highly pressurised activity, particularly so when it takes place in a competitive tendering scenario. Most people who invite competitive tenders want them returned in a hurry, for whatever reason.

Figures 10.4 and 10.5 illustrate a typical project involving a competitive tendering process. Out of, say a 30-month project period for the design and construction of a new building, perhaps one month will be allocated to the process in respect of building work, and if there is a tender competition for the appointment of consultant designers, they will be given about the same but at the start.

Tendering for consultancy work is generally straightforward, as the main resources used are time and money, expressed as estimated hours for various levels of staff, multiplied by an appropriate hourly rate. The difficult part lies in understanding exactly what is wanted by the client, and that is always a

Action	Week 1							Week 2							Week 3							Week 4						
	1	2	3	4	5	6	7	8	9	10	11	12	13	14	15	16	17	18	19	20	21	22	23	24	25	26	27	28
Receive tender	●																											
Direct work	→	→	→	→	→	→	→	→	→	→	→	→	→	→	→	→	→	→	→	→	→							
Subcontractors				→	→	→	→	→	→	→	→	→	→	→	→	→	→	→	→	→	→							
Suppliers				→	→	→	→	→	→	→	→	→	→	→	→	→	→	→	→	→	→							
Temporary works	→	→	→	→	→	→	→	→	→	→	→	→	→	→	→	→	→	→	→	→	→							
Overheads	→	→	→	→	→	→	→	→	→	→	→	→	→	→	→	→	→	→	→	→	→							
Returned prices															▓	▓	▓	▓	▓	▓	▓							
Net tender price																						●	●					
Adjudicate tender																							●					
Submit bid																								●				
Planning	→	→	→	→	→	→	→	→	→	→	→	→	→	→	→	→	→	→	→	→	→	→	→	→	→	→		
Outline programme	→	→	→	→	→	→	→	→	→	→	→	→	→	→	→	→	→	→	→	→	→	→	→	→	→	→		
H&S plan?																												
Risk assessment?																												

Figure 10.4 Tender process in construction

function of the design brief or documentation accompanying the tender enquiry. The pricing process is simple, it is the informing and understanding process that is difficult. Three- to four-week tender periods are generally adequate for consultant tendering unless there is a design competition attaching.

Tendering for construction work is a much more complex process which is usually preceded by pre-qualification procedures, designing and perhaps demolition and enabling works on site (Figure 10.5).

Upon receipt of a tender enquiry, tendering main contractors will usually examine the documents and break them down in the following way:

- work to be undertaken direct;
- subcontractor packages;
- supplier packages;
- temporary works;
- site-based overheads (preliminaries).

In a four-week period, this will take about a week, including the assembly and despatch of the subcontractor and supplier packages. It is not unusual for many projects to be entirely subcontracted in this way, with the contractor planning to do little direct work on site himself. Subcontractors and suppliers will be asked to return their tenders by the end of week three at the latest, and until they do, nothing much will happen in the offices of tendering main contractors. All that can happen is the pricing of any direct works, and perhaps

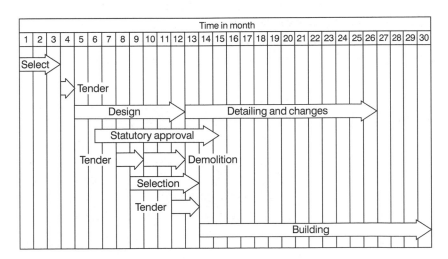

Figure 10.5 Competitive tendering

some temporary works design and pricing. There is no point in pricing the site-based overheads (preliminaries) until subcontractor and supplier prices are returned, because they may contain qualifications or exclusions which impact on site overhead pricing.

By the start of week four, fewer than half of subcontractors and suppliers invited to tender will have actually done so. Clerical staff in contractors' offices will start progress chasing. By about Wednesday of week four (day 24 out of 28) estimators will get busy. Over the next two days, they will cobble together a net tender total which will be their best wild guess of the prices to be paid to directly employed resources, subcontractors and suppliers, to construct, complete and maintain, and if appropriate, to design, the works described in the tender documentation.

They will then submit this, sometime late on Thursday or even early Friday (days 25 and 26) to an adjudication board comprising a couple of directors (so that a single director can never be held solely responsible) who will decide whether to raise or lower the estimator's best wild guess, and how much or little to add to the resulting sum in respect of site and head office overheads and profit.

The effective work on this tender will have been done in about two days, and in the offices of subcontractors and suppliers. At best it represents the tenderer's best guess of the price he thinks he needs to quote to win the work, if he wants it. It is a procedure to procure a price, and it bears little resemblance to the competition between firms of similar skills and capacity based on the cost of carrying out the works, plus a reasonable mark-up, that is it is meant to be.

That is one of the reasons why Sir John Egan and the Construction Taskforce in their report 'Rethinking Construction' (July 1998) recommended greatly reduced reliance on competitive tendering, pointing out that whilst this may go against the grain, especially for the public sector, 'it is vital that a way is found to modify the process so that tendering is reduced'.

10.5 How does the pricing process affect safety?

I suggest that, notwithstanding the vagaries and inaccuracies of the commercial aspects of tendering, there is an even more compelling reason for moving away from it. If my description of the process is accurate, and I believe it is, then it is simply not possible for tendering main contractors and subcontractors to demonstrate the adequate resources they have allocated or will allocate to managing health and safety, in construction projects to the standards envisaged by the CDM Regulations, and in particular Regulations 15, 16, 17, 18 and 19, and for clients Regulations 9 and 10.

Egan's taskforce felt that quantitive performance targets and open book accounting, together with demanding selection arrangements and proper auditing were the way forward. This concept is known as 'target cost contracting' and it is not new, but has previously fallen foul of government policy in the form of compulsory competitive tendering, and the lowest price mentality of many clients.

The essence of target cost contracting is that the parties to the contract agree ways of establishing target cost for each work section as work proceeds. The Chartered Institute of Building CoEP methods, described earlier in this chapter, are entirely suited to that process and are conducive to open book accounting, to facilitating changes and to audit. It also allows contractors to demonstrate that adequate resources have been allowed.

As an aside, there are several standard forms of contract available to implement this method including the Engineering and Construction Contract, which has the added advantage of a family of back-to-back consultants and subcontractor agreements which can be used without amendment.

A final point on competitive tendering: its cost. The following is an estimate of the total cost falling upon the supply side in tendering for, say, a new primary school, using a JCT80, With Contractor Design Portion Supplement Contract. Assume a green field site, and an estimated project value of about £3 million and a fairly traditional form of construction.

Cost for preparing one competitive tender

Main contractor's in-house costs:

		£	
Technical staff	300 hours @ £30 per hour	9000.00	
Administrative staff	300 hours @ £15 per hour	4500.00	
Photocopying, postage, telephone		1500.00	£15 000.00

Subcontractors and suppliers:

20 subcontractor packages	× 6 = 120		
20 supplier packages	× 6 = 120		
10 plant packages	× 6 = 60		
Total packages	300		
Half do not reply	(150)		
Total returned packages	150 ×	£2000.00	£300 000.00

Cost to subcontractor and suppliers:

	£
40 hours @ £30 per hour	1200.00
40 hours @ £15 per hour	600.00
Photocopying, postage, telephone	200.00
	£2 000.00

Cost per tendering main contractor	£315 000.00
Six tendering contractors	× 6
Cost falling upon the supply chain	£1 890 000.00

An astonishingly high cost approaching £2 million falls on the supply side every time a client competitively tenders a project such as this. These costs do not stay with the supply side, they are inevitably passed back to clients through the mark-ups applied for overheads and profit.

The waste inherent in this process is staggering. Only one tender can be successful, so the other five are thrown away. The cost of the successful tender is not recovered directly through the rates and prices charged in the tender sum, so arguably the entire £2 million is wasted in the sense that it makes no direct contribution to the process. Tendering is simply a very expensive, unreliable and wasteful selection device based on the unsupportable notion that the lowest price is also the best tender.

It also, in my view, increases both the magnitude of risk and its probability in terms of demonstrating adequate resources under the CDM Regulations.

Egan's taskforce's view of the move away from tendering is that it will mean radical changes in the culture of construction, resulting in fewer but bigger winners among construction firms. In my view the radical culture change that is needed is not amongst contractors, because I believe that they will willingly give up competitive tendering, but amongst clients because it is clients who insist upon competitive tendering. The foreword to the CDM Regulations also mentions that a radical change in culture is needed in construction.

Dutch auctioning

One of the features of competitive tendering in construction projects is the extensive use of 'dutch auctioning'. Clients do it to consultants and main contractors, the latter do it to specialists and suppliers. Most deny doing it, or dress it up under some hazy theory about 'markets'. Basically the procedure involves the receiver of a tender telling the bidder 'knock 5 per cent off your tender price and I will give you the job'. Purchasing professionals describe the technique as 'adversarial leverage'.

In the health and safety sense the dilemma is this. If the bidder has submitted a bid which meets the established criteria, which must include compliance with health safety laws including CDM, then what does he knock the 5 per cent off? Is it unit rates and prices for the work prior to the inclusion of health and safety resources? In which case, he cannot do the work for the prices he is now quoting and will lose money. Or is it the adequate resources? In which case, they are now no longer adequate, and he risks breaking the law.

Pricing and work sequences

Experienced estimators know that it is rarely possible to price individual items of work without reference to or in consideration of other activities adjacent to them in terms of time, sequence or location. For example, the construction of manholes cannot be considered in isolation to the drain runs between them. The conventional wisdom with drainage systems is that they must be built 'deep to shallow'. If they are built the other way round, and there is heavy rainfall, the drainage system will work, and the deep drain trenches will flood from the shallow end. It will then be necessary to de-water them, thus adding to the cost of the operation (Figure 10.6).

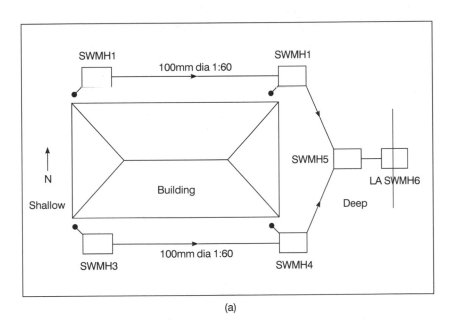

(a)

Operation	Week 1							Week 2							Week 3							Week 4						
	M	T	W	T	F	S	S	M	T	W	T	F	S	S	M	T	W	T	F	S	S	M	T	W	T	F	S	S
Manholes – 5 No	▓	▓	▓																									
LASWMH6 – SWMH5								▓	▓	▓																		
SWMH5–2 Drains															▓	▓	▓											
SWMH2–1 Drains															▓	▓	▓											
SWMH5–4 Drains																						▓	▓	▓				
SWMH4–3 Drains																						▓	▓	▓				
Test the system																									▓	▓		
Backfill 3 Grade																									▓	▓		
Start brickwork													········	North	elevation	········						▓	▓	▓				
Manhole covers																									▓	▓		

(b)

Figure 10.6 Pricing sequence

Figure 10.6 depicts a simple surface water drainage system. The deepest point on the system is at local authority surface water manhole 6, the two shallowest points are at surface water manholes 1 and 3, and all other points on the system fall at 1:60 from shallow to deep. The drainpipes are 100 mm diameter.

The programme in Figure 10.6 shows that the contractor intends to construct all five manholes in week 1 (LA SWMH6 exists).

- Once the manholes are in to the correct depths, but without their permanent covers, the drain runs are easy to establish.
- In week 2, the contractor will install drain runs LA SWMH6–SWMH5, keeping the deep to shallow methodology.
- In week 3, the contractor installs SWMH5–SWMH2, and SWMH2–SWMH1, thus freeing the north and east elevations for the follow-on operation in week 4 of external wall brickwork and blockwork.
- Also in week 4, drainage resources transfer to the drain runs SWMH5–SWMH4 and SWMH4–SWMH3 along the south elevation.
- When the whole system is in place it can be tested and backfilled, and the permanent manhole covers put in place. This is programmed for the weekend of week 4.

There is a work sequence upon which the contractor will base his tender price. There is some in-built flexibility which would not affect the validity of the prices. For example, the contractor could reverse the north/south sequence, and install the runs SWMH5–SWMH4–SWMH3 before SWMH5–SWMH2–SWMH1.

However, if he did this and still attempted to start north elevation brickwork in week 4, this would significantly increase the risk factor to brick layers and probably the drainage gang. They would be getting in each other's way and the vehicle and plant movements associated with their activities could become high risk.

Hopefully, the following points are demonstrated:

- Unit rates and prices are expected to stand alone.
- However, they are nearly always calculated in the context of a work sequence.
- Work sequences are usually interdependent.
- Work sequences cannot normally be changed without impacting on each other.
- Changing the work sequence usually increases risk in terms of likelihood or severity or both.

Final words on pricing and tendering

Pricing and tendering are not the same. They are two distinct and separate but related processes.

The pricing process is the stage at which adequate resources are built into rates, outputs and prices. Any pricing system should be configured to take into account risk assessments and method statements based upon identified work sequences.

Tendering is no longer compulsory in the public sector, and never has been in the private sector. In the UK construction industry in the 1990s, it is a discredited process, the use of which is discouraged by the construction taskforce and some large client bodies. Tendering presents two main threats to good health and safety management, namely the submission of tender prices at or below a level at which the work can be done profitably and safely, and post-tender dutch auctioning. This is an unethical procedure whereby already low prices are driven down even further, thus encouraging dutyholders to compromise on the provision of allocated adequate resources.

Appropriate use of alternatives to traditional tendering methods based on bills of quantities and adversarial forms of contract can be helpful in avoiding these pitfalls. For example, the use of the target cost options of the Engineering and Construction Contract or GC Works series can assist the parties to negotiate competitively without compromising on safety.

10.6 Pricing and procurement routes

The six different procurement routes were examined in detail in Chapter 5, and the impact of each on health and safety management strategies was explained. The choice of procurement route also acts as an indicator of the pricing strategy, and will configure the way in which project prices are compiled.

The traditional route

The traditional route contemplates a complete design prepared by consultants which is then usually put out to tender to contractors. As an aside, there is nothing in the main standard forms of contract (JCT and ICE) that calls for a compulsory tender competition. It is both possible and permissible to negotiate a price.

Upon receipt of a complete design, contractors are instructed to price 'the design' which includes drawings, design details, specifications and can include a bill of quantities. The bill serves as a vehicle to enable a tender price to be assembled, and the Chartered Institute of Building CoEP has been specifically designed to facilitate the pricing of projects from bills of quantities prepared in accordance with the current edition of the 'Standard Method of Measurement' published by the Royal Institute of Chartered Surveyors.

Accordingly a complete, or nearly complete design facilitates accurate and detailed pricing of the project in accordance with the CoEP, and for that reason, it also facilitates a detailed address of the adequate resources mentioned in CDM Regulation 9.

However, there are some 28 recognised variants of the traditional route where bills of quantities are not used. These include drawings and specification, partial contractor design, performance specified work, approximate quantities and so on, where the contractor is invited to submit a fixed-price lump sum for the project. The degree of pricing detail available at the time of tender invariably decreases the further the project moves away from the notion of a complete design. The address of adequate resources is sensitive to the decreasing pricing detail and it becomes more difficult to estimate adequate resources.

Design and build

The tender document here is the employer's rrquirements (or similar) and the contractor responds with the contractor's proposals. The intention is to procure a fixed price lump sum, accompanied by a schedule explaining how it is calculated. The CoEP methodology can be applied to the calculation of the tender price, but the default position is that design and build contractors will subcontract most if not all of the works. Consequently they will use prices procured from subcontractors and there may in fact be no attempt to estimate from first principles. This could give difficulty in demonstrating compliance with CDM Regulation 9 and adequate resources.

Design and manage

The same point applies to design and manage. The tender price will usually comprise subcontract and supply quotations and it is usually necessary to drill down into them to ascertain adequate resources. The route is also noted for the 'dutch auctioning' of supply prices.

Management contracting

In this route the client has a contract with consultants for partial amounts of design, and with a management contractor for completion of the design and provision of construction work through a series of work package contracts with specialist suppliers and subcontractors.

One of the claimed benefits of the route is that it allows an early start on site in the absence of a complete design. Designing and building can be overlapped. However, price certainty cannot be achieved until the last works package has been let, which can be quite late in the construction sequence.

This route gives obvious difficulties in identifying adequate resources in works packages, and further difficulties can be experienced at the interfaces of packages and with the management contract package itself. Management contractors usually strongly reject this point, claiming that their management skills are strong enough to overcome the difficulties. There are projects where this is the case, but when contracts go wrong they default to a letter box operation and in the writers experience, health and safety management systems can break down.

Construction management

Here, the client is in direct contract with all of the consultants and package contractors needed to deliver the project, and the construction manager provides a separate package co-ordinating and managing the project and the supply chain.

The route has the same potential difficulties as management contracting, arising from the same cause. Because the works packages are assembled sequentially, it may be quite late in the project before the principal contractor can know all of the adequate resource requirements. Also the developing design requires on-going vigilance by planning supervisors, and a continuing address of CDM Regulation 13 by designers.

Engineer procure construct

Otherwise known as 'turnkey', this route contemplates the complete designing, building and commissioning of a process which includes construction. The example quoted in Chapter 5 was for a combined heat and power plant.

Usually the client writes a description of the service to be purchased, and contractors respond with their outline proposals and a price. The proposals are outline at this stage because bidding contractors will not submit detailed proposals until they know they have secured the contract.

Consequently a detailed address of adequate resources at tender stage is usually not possible, to the standard envisaged by the Regulations. For this reason quoted prices tend to be of the approximate estimate or best wild guess variety. The principles of the CoEP can be applied to turnkey projects but the onerous nature of bidding competition usually mitigates against this.

Many turnkey projects are subjected to HAZOP procedures which are applied as package contracts are developed, so there is a mechanism that can be applied in many cases. However, the route can suffer from similar practical difficulties to those encountered in works packages developed in the management contracting and construction management routes.

10.7 Pricing and contract strategies

Contract strategies for managing health and safety in the standard forms of contract are examined in detail in Chapter 9. The impact of the contract strategy on the pricing of projects can be found in the payment mechanisms set out in the contract matrix. The broad options are set out at Section 10.3 of this chapter and are repeated below, namely:

- priced contract with activity schedule;
- priced contract with bill of quantities;
- target contract with activity schedule;
- target contract with bill of quantities;
- cost-reimbursable contract;
- management contract.

In general terms, bills of quantities allow for greater detail in pricing, and facilitate the allocation of adequate resources at the time of pricing or tender.

Lump sum prices generally mean less detail in both pricing and adequate resources at the time of tender.

Cost-reimbursement contracts probably afford a greater degree of certainty that adequate resources have been allocated.

Management contracts tend to be simpler to deal with in terms of resources as they rarely contain significant plant or material elements.

In unamended form, all of the standard forms contain provisions for dealing with changes in design and in work content which allow price and adequate resource to be varied.

10.8 Future issues affecting price

New legislation introduced in 1996 and 1998 will impact upon the way in which construction activity is priced. The main regulations are:

- Health and Safety (Safety Signs and Signals) Regulations 1996;
- Health and Safety (Consultation with Employees) Regulations 1996;

- The Working Time Regulations 1998;
- The Provision and Use of Work Equipment Regulations 1998;
- The Lifting Operations and Lifting Equipment Regulations 1998.

Health and Safety (Safety Signs and Signals) Regulations 1996

These Regulations require employers to use safety signs in the form of a pictogram instead of text only wherever there is a risk to health and safety which cannot be avoided or controlled by other means.

Health and Safety (Consultation with Employees) Regulations 1996

These regulations extend employee consultation rights to non-unionized workplaces. This is particularly important in construction which makes extensive use of subcontracting and where the influence of unions has been curtailed in recent years.

The Working Time Regulations 1998

The Working Time Regulations limit the number of hours in any given week which employees can be required to work, and also give employees the right to annual paid leave. They came into force on 1 October 1998. Clearly the provisions of these regulations will impact upon the calculation of unit rates and prices, particularly where extensive use of overtime or out of normal hours working is envisaged, or is an essential part of any construction activity.

The regulations apply to:

- workers over the minimum school leaving age;
- part-time workers, freelancers, temporary staff and agency staff.

The main provisions of the regulations are:

- from 1999 onwards, four weeks paid annual leave;
- eleven consecutive rest hours in any 24-hour period;
- twenty minute break where more than six hours is worked;
- forty-eight hour average working week, unless the employee chooses to work more;
- eight hour average in any 24 hour period for night workers;
- health assessment entitlement for night workers;
- employers must keep records of hours worked by employees.

There are a number of 'derogations' – circumstances where the regulations do not apply, mainly arising from circumstances where work time cannot be accurately pre-determined or measured.

The entitlements under the regulations are enforced by employment tribunals, whereas the time limits are enforced by the Health and Safety Executive and local authorities, and can therefore be the subject of criminal proceedings against employers.

Provision and Use of Work Equipment Regulations 1998

PUWER came into force on 5 December 1998, and is accompanied by an approved code of practice. PUWER 1992 is completely replaced, but its provisions are carried forward into the new regulations and are expanded.

PUWER 1998 applies to work equipment as follows:

- toolbox tools – hammers, hand saws, etc.;
- single machinery – circular saws to dump trucks;
- apparatus – theodolites, levels etc.;
- Lifting equipment – cranes, hoists, lifts, etc.;
- installations – groups of machines such as a concrete batching plant.

The main requirements of PUWER 1998 include:

- further guidance in respect of specific risks in the approved code of practice;
- dutyholder applications are extended to include people who control the way in which work equipment is used, including plant hire companies;
- inspection requirements in respect of work equipment where significant risk could occur from incorrect installation, movement, relocation, deterioration or exceptional but foreseeable circumstances;
- record keeping of inspections.

PUWER 1998 now applies to power presses and woodworking equipment and repeals and replaces the old laws applying to them.

PUWER 1998 will impact directly on all dutyholders under the CDM Regulations, starting with designers. Any work equipment, such as cleaning and access equipment that they design into the permanent works, or process equipment such as production line equipment, will almost certainly come within the scope of PUWER 1998, and CDM Regulation 13. It will be necessary for designers to show how they have applied the hierarchy of risk control to some if not all aspects of their design, and to be in a position to provide relevant information for inclusion in the health and safety file.

Principal contractors and contractors will be expected to comply directly with PUWER 1998 in respect of all plant and equipment that they use during the course of their work activities on site.

Planning supervisors must be in a position to understand how PUWER 1998 applies to any project to which they are appointed.

Clients must be able to comply with PUWER 1998 once they take possession of the completed structure in any project.

The cost of compliance with PUWER 1998 must therefore be included in rates, prices and tender sums and be demonstrable in the form of adequate resources. These must address:

- minimizing risks from overturning;
- provision of breaking and stopping equipment;
- prevention of start-up by unauthorized persons;
- ensuring drivers have adequate fields of vision.

Similarly the cost of compliance with the next set of regulations must also be built into rates, prices and tender sums.

The Lifting Operations and Lifting Equipment Regulations 1998

The main requirements of LOLER is for employers to carry out a risk assessment of planned lifting operations taking into account the following factors:

- the load being lifted;
- weight, shape and centre of gravity;
- availability of lifting points;
- the way the lifting equipment is to be used;
- environment in which lifting is to take place;
- available personnel, and degree of involvement.

Employees' risk assessment must take into account selection and use of lifting equipment and any necessary staff training. Lifting equipment, which covers a wide range of applications such as tower cranes, mobile cranes, excavators used as lifting devices, fork lift trucks and the like, must be of adequate strength and stability and risks arising from positioning, installing and dismantling must be considered. Safe work loads must be clearly marked.

Lifting accessories such as slings, hooks and shackles must be examined every six months and all other equipment must be inspected every twelve months.

Measures must be taken to prevent overturning and overloading and there are special requirements involving equipment used to lift people.

LOLER 1998 repeals and replaces all of the old fragmented Regulations applying to lifting operations, enshrined in construction in the Construction (Lifting Operations) Regulations 1966. In most cases, LOLER 1998 will need to be considered at the same time as the Construction (Health Safety and Welfare) Regulations 1996 if any lifting operations are contemplated on site.

Also if the permanent work design includes any lifting equipment and designers, planning supervisors and clients will need to take LOLER 1998 into account.

COSHH 1998

COSHH 1998 is one of a number of sets of regulations which were revised in that year. Under the new regulations employers are required to protect their staff from harmful substances. A number of substances are now prohibited, namely those containing eight chlorinated solvents including carbon tetra-chloride and chloroform. These are often used in industrial solvents or cleaning compounds used for degreasing and can release harmful vapours into the atmosphere, particularly if they catch fire. The writer was involved in the aftermath of an accidental fire in 1988 in an engineering works in the East Midlands. A 200-gallon tank of degreasant had caught fire and the direct fire damage repair bill amounted to about £50 000. However, under combustion, the solvent broke down into hydrochloric acid vapour and phosgene (mustard gas). Upon expert examination about three weeks after the fire, the entire outer shell of the building was condemned because of the presence of large quantities of these toxic materials. It was eventually necessary to replace the entire outer shell of the building at a direct cost of about £680,000 and business interruption costs of about £1 million.

The Control of Lead at Work Regulations 1998

These regulations reduce the lead level concentrations at which employers must prevent employees from working with lead.

Gas Safety (Installation and Use) Regulations 1998

These are intended to safeguard the public from problems and risks which might arise from faulty gas appliances. They are accompanied by an approved code of practice and replace the Gas Safety Regulations 1994.

CE markings

The Construction Products Directive sets out requirements for European Technical Approvals (ETAs) which are intended to apply to most products used in construction throughout the UK and the EU. This requires manufacturers to develop appropriate attestation of conformity procedures in respect of their products. So far this has proven difficult to do because of the absence of suitable ETA accreditation bodies. One of the few in existence is the British Board of Agrément, but more are being developed.

A product that is legally CE marked cannot be refused entry to any market within the European economic area. It is expected that the provisions of the directive will be implemented by a new set of domestic regulations in the UK in the near future.

10.9 Adversarial versus co-operative contracts

The payment mechanisms in adversarial and co-operative contracts can have different impacts on the management of health and safety within the project.

Adversarial contracts include most JCT and ICE standard forms, and they contain provisions within them that reflect the English legal system's way of doing things, which is adversarial. It is important not to construe the word in its pejorative sense here. It simply means that, to the extent that one party to a contract believes that the other is in breach of any term or condition of the contract, it is the responsibility of the complaining party to prove its allegation to the defending party.

Accordingly the adversarial contracts all contain provisions which require the plaintiff (usually the contractor) to issue notices to the defendant (usually the employer) specifying each and every breach, detailing causes and effects arising from each breach and particularizing the loss and expense flowing from the breach. The defendant is entitled to know the case against him or her.

The notices to be issued by the complaining party serve two purposes. First, to preserve the issuing party's entitlement to any extra time or money, and second, to give the receiving party the opportunity to do something about the complaint either before it becomes breach, or in the alternative to mitigate the effects of any existing breach.

The procedures surrounding the issuing and receiving of notices have become known under the all embracing name of 'claims' and there is an entire mini-industry within the construction process that deals with claims. It is a sad reflection on the industry that it spends far more (about £750 million per

303

annum) on dealing with 'claims' than it does on research and development (about £250 million in a typical year).

The major difficulty with claims can be found in two main areas, namely:

- the role of the contract administrator/engineer;
- proving breach and its consequences.

The notices of alleged breach in a JCT or ICE contract are sent by the contractor not directly to the employer, instead they go to the contract administrator (in a JCT contract) or the engineer (in an ICE contract). He must then ascertain the facts, the provisions of the contract and the law behind the contractor's notice or claim, and he acts in a quasi arbitral role to determine any entitlements in terms of extra time or money, or both, due to the contractor and arising out of or from the alleged breach.

However, the contract administrator/engineer may also fulfil two other roles, namely designer of the permanent works and client's agent, and in those capacities it is commonplace for contractors to allege that the breach has occurred as a result of some action, or failure to act, by the designer or client's agent.

In other words, the contract administrator/engineer is expected to act as impartially as an arbitrator or a judge in awarding the contractor extra time and money because of his own possible failure to discharge one of his other roles to the required standard. This leads directly to the second difficulty.

Alleging breach of contract is one thing, proving it is quite another. The same applies to actually demonstrating alleged cause, effect, loss and expense. The contract administrator/engineer cannot make any award of time or money until he has been provided with the required proof by the contractor. The outcome of these procedures is usually conflict which can lead to dispute, but also to protracted arguments about time and money. Basically the complaining party (usually the contractor) does not get paid until he has proven his case to the reasonable satisfaction of the contract administrator/engineer acting as quasi arbitrator. In the meanwhile the contractor is obliged to carry out the work which is the subject of his claim or claims to the same standard as the rest of the works which are not in dispute.

Clearly on contracts where there are number of claims, protracted delay in dealing with the claims for extra time and money can lead contractors into unsafe work practices for a number of different reasons. These can include lack of adequate information, lack of funds to pay for adequate resources and pressure on time in attempting to meet unextended completion dates.

Co-operative contracts are so-called because they are said to force the parties to behave in a co-operative way in terms of dealing with any potential breaches of contract. The Engineering Contract and the GC Works series (1998) are examples of co-operative contracts and, by way of example, the Engineering Contract works in the following way.

First, the role of contract administrator is split into three, that is, project manager, quality supervisor and adjudicator. Each is appointed separately.

The project manager's role is then straightforward. He or she is there to look after the employer's interests and ensure that his objectives are met.

The project manager and the contractor are then required to work together to hold regular 'early warning meetings' and by reference to the design, the contract documents and the contractor's programme and any proposed change orders, to identify any compensation events that might give rise to a claim or claims.

They are then obliged to agree any time and money entitlements flowing from the compensation event before the work is actually put in hand. This enables the employer to know the case against him before it happens, and the contractor to know his entitlements before he does the job.

It is a lot easier to describe the mechanism than it is to actually make it work. Nevertheless many people prefer this system because it enables adequate resources to be built in to the compensation event payment mechanisms in terms of time and money, and for those resources to be demonstrated and deployed. It also facilitates development of health and safety plans, for the same reason.

10.10 The Scheme for Construction Contracts (England and Wales) Regulations 1998

The Scheme deals, *inter alia*, with adjudication pay-when and pay-if-paid clauses in contracts and came into force on 1 May 1998. The regulations are made under the Construction, Housing Grants and Regeneration Act 1996, and on the face of it they have nothing to do with health and safety management on building sites, or the pricing of construction work. However, examination of some of the provisions of the Act and the scheme reveals otherwise.

The usual reason for contractors and subcontractors threatening to walk away from a job is non-payment, either of certified sums properly due, or amounts allegedly due under the heading of claims. Failure to honour payment certificates usually constitutes breach of contract; failure to deal with claims in accordance with the contract terms can also amount to breach.

There is no automatic right under English law to suspend work, or pull off site, in the event of breach of contract. The usual remedy given to injured parties is 'determination' and detailed procedures for this are usually included in the contract. It is not an easy option.

The Scheme for Construction Contracts (England and Wales) Regulations 1998 is intended to deal with several specific thorny problems that have plagued the construction industry for decades, *inter alia*:

- pay-when and pay-if-paid clauses in contracts;
- quick dispute resolution;
- the right to suspend work.

It is standard practice to include clauses in contracts, especially subcontracts, to the effect that:

> Payment will be made to the subcontractor three days after receipt by the main contractor of any sums properly due . . .

This is a 'pay-when' clause.

> Payment will be made to the subcontractor as and when monies are received by the main contractor from the employer.

This is a 'pay-if' clause.

Such clauses are now rendered ineffective by the provisions of the scheme. They are not banned; instead if, say a main contractor attempts to impose a pay-when/pay-if clause on a subcontractor, the subcontractor can dispute the matter and seek a quick resolution to the issue through adjudication.

Adjudication has been described as rough justice. Adjudicators differ from arbitrators in so far as they operate under the rules of expert witness and their decisions, provided they are given in good faith, cannot normally be challenged on appeal. Adjudication is now a right given to all parties to construction contracts, oral and written, and is implied into them if not expressly written.

Adjudication operates to a quick time scale. If the subcontractor mentioned above believes that pay-when/pay-if-paid is being imposed on him, or if he believes he is not being properly paid in accordance with the terms of the contract, he has the right to ask the main contractor to agree to the appointment of an adjudicator.

The main contractor has seven days in which to agree. If he ignores the request or prevaricates, the adjudication runs anyway and the adjudicator gives his decision within 28 days unless the parties agree to extend the period. It is usually not in the complaining party's interests to do that.

So at best, 36 days after a dispute has crystallized it can now be decided by an adjudicator. If the defending party fails to abide by the adjudicator's decision and award, the injured party now has the right to suspend work on site until compliance is achieved.

The consequences of suspension on health and safety management on site could be grave. Many examples spring to mind and a few are given below.

- scaffolding subcontractors;
- tower crane hire companies;
- access equipment hirers;
- groundworks subcontractors.

If any of these were to suspend work it would almost certainly paralyse the project.

It is emphasized that all parties to construction contracts, oral and written, now have the statutory right to implement the provisions of the scheme and can suspend work if adjudicator's decisions are not implemented. Consequently it is in all parties' best interests to resolve disputes, particularly those involving pricing and payment, quickly and to each other's mutual satisfaction. A suspension of work by one trade will almost certainly compromise the health and safety management standards of all of the trades on site.

10.11 The Building Regulations

The building control system in England and Wales was radically changed in 1985, and the Building Regulations currently in force commenced on 1 June 1992, were amended in 1994 and are also updated regularly.

The Building Regulations are made with three broad objectives in mind.

- energy conservation;
- prevention of unnecessary waste, contamination, consumption or misuse of water;
- ensuring the health, safety, welfare and convenience of members of the public in or affected by building in the environments.

It is sometimes forgotten that the Building Regulations are probably the most comprehensive set of regulations affecting all sectors of construction and that they have a strong central health and safety component.

The regulations themselves are quite short, but they are accompanied by a series of 14 approved documents and some non-statutory guidance which collectively comprise a voluminous amount of technical detail, much of which is indexed and referenced to other non-statutory material including British Standards and codes of practice. Building Regulations are statute law and must be complied with. The cost of compliance must be built in to the rates and prices charged by and paid to consultants, contractors and specialists when carrying out construction work.

The Building Regulations are enforced by local authorities through their development and planning control arms. The regulations are far-reaching and comprehensive and apply to designing, constructing, new build and maintenance, refurbishment and demolition, but they are not usually thought of as health and safety laws in the same way as the Management Regulations, CDM and the like. Instead they can be thought of as a codified set of minimum performance and technical standards with an in-built health, safety and welfare component.

10.12 Chapter summary and conclusions

The pricing of construction must embrace the adequate resources needed within the overall health and safety management process in projects.

Adequate resources includes plant, material, technical facilities, trained personnel and time needed to fulfil the obligations of dutyholders.

Risk assessment is an important part of the pricing process, which may or may not take place in a competitive tendering environment.

Competitive tendering and adversarialism can present barriers to successful health and safety management, and careful attention to the choice of procurement route and contract strategy can help to overcome some of the more obvious pitfalls of the tendering process.

Legislation is made regularly and frequently, and a number of regulations which came into force in 1996 and 1998 will affect pricing, both before and during the construction stages of projects.

The Building Regulations remain perhaps the most comprehensive set of regulations affecting all aspects of health and safety management in construction.

Epilogue

This book started with a glance at some elementary 'building regulations' laid down in the book of Deuteronomy and finishes at a modern set a couple of millennia or more later. I agree with Sir Bernard Rimmer, a well-known commentator on the UK construction industry and chairman of Slough Estates, who is reportedly fond of saying, 'There is nothing new in construction.'

The Construction Taskforce under Sir John Egan, in its July 1998 report to the deputy prime minister 'Rethinking Construction', set the construction industry a target of a reduction in the number of reportable accidents by 20 per cent year on year. Egan noted how some clients and construction companies have achieved a reduction in reportable accidents of 50–60 per cent over a time span of two years or less.

The taskforce pointed out that most accidents on construction sites occurred, in its experience, through lack of training or from people working out of process, a theme that runs right through this book.

It will be very interesting to see, over the next few years, whether or not the construction industry can meet Egan's challenge. I believe we will, and in my new job at Interior plc, I am going to try to get reportable accidents down to zero, and keep them there.

Appendix 1:
Health and safety policy template and outline of a safe work system

This is prepared as a template intended to be of assistance to a broad range of employers and will need specific adaptation to each employer's circumstances.

Definitions and explanations of terms used

Competent person: Someone with sufficient training, experience, knowledge and qualifications to do the job in hand. Competence varies with the complexity of the problems or situation being addressed. Most often it is understood to mean possessing such theoretical and practical knowledge as will enable conclusions to be drawn and/or actions to be taken.

Foreseeable: For an occurrence to be foreseeable it is not enough that a remote possibility of injury exists. There must be sufficient probability for a reasonable person to anticipate it happening.

Hazard: The potential to cause harm including ill health and injury, damage to property, plant, products or the environment, production losses or increased liabilities.

Measuring: The collection of information about the implementation and effectiveness of plant and standards involving a variety of checking or monitoring activities.

Monitoring: A structured audit process of collecting independent information on the efficiency, effectiveness and reliability of the safety management system and drawing up plans for corrective action.

Organization: A general term to describe the responsibilities and relationships between individuals which forms the social environment in which work takes place. Organizing is regarded as the process of designing and establishing those responsibilities and relationships.

Planning: The process by which the objectives and methods of implementing the health and safety policy are decided.

Policy: The general intentions, approach and objectives of an organization and the criteria and principles on which actions and responses are based.

Reviewing: Activities involving judgements about performance and decisions about improving performance. Reviewing is based on information from measuring and monitoring activities.

Risk: The likelihood of the potential to cause harm being realized.

Shared premises: Means one or more of the following:

- under the same roof;
- sharing the same access/egress;
- sharing common areas of the same building or site;
- sharing the same site;

Undertaking: The employer's business as a whole which can include more than one premises, workplace or site under the employer's control.

Company name			
Address			
Signatory			
Signature		Date	
Job title			
Nature of firm's business	Private sector	Leisure	
	Public sector	Service	
	Industrial	Culture	
	Agricultural	Educational	
	Commercial	Administration	
Size of business at this location	Site area		
	Number of buildings		
	Total building area		
	Total number of employees		
	Single-storey buildings		
	Multi-storey buildings		
	Other types		
Description of work activities and products made at this location			

Regulation 4

Arrangements for planning, organization, control monitoring and review procedures

Regulation 4

Planning Purchase and provision of work equipment *Name of person responsible*	Management arrangements
Organization Information and co-operation arrangements Company's health and safety *Name of person responsible*	Management arrangements
Control Decision making and implementation procedures *Name of person responsible*	Management arrangements
Monitor and review Compare planned/actual and implement improvements *Name of person responsible*	Management arrangements

Health and safety assistance

Regulation 6

The following is a record of the competent people available both internally and from external specialists.

External	
Name of company Address Tel. and fax Contact name	The company employs the services of XYZ plc to provide it with specific advice on construction-related matters as follows: List arrangements:

Internal	Job title	Training record	Competence levels
Name			
Name			
Name			
Name			

Hazard identification and risk assessment procedure

Regulation 3

Alternative 1: Workplaces

The workplace is divided into the following hazard identification (hazardident) zones:

Zone 1 Zone 2 Zone 3 Zone 4 Zone 5	This approach is best suited to non-construction or complex construction related activities.

Form JRK 1	**Hazard identification form** Hazards are identified in each zone on Form JRK1
Form JRK 2	**Risk assessment form** The hazards identified on JRK 1 are risk assessed on Form JRK2 using the company's standard risk assessment procedure
Form JRK 3	**Preventive and protective measures** The risks assessed on JRK2 will reveal the preventative and protective measures necessary to comply with relevant regulations and will be recorded on JRK3.
Form JRK 4	**Safe work system** The risk assessment procedure will reveal work activities which require the design and implementation of a safe work system which will be recorded on JRK4.

Hazard identification and risk assessment procedure

Regulation 3

Alternative 2: Work activities

The construction activities are divided into the following hazardident activities:

Demolition Site clearance Piling Excavation Foundations Floor slabs	This approach is best suited to standard construction products or sequential flow-related production activities.

Form JRK 1	**Hazard identification form** Hazards are identified in each zone on Form JRK1
Form JRK 2	**Risk assessment form** The hazards identified on JRK1 are risk assessed on Form JRK2 using the company's standard risk assessment procedure
Form JRK 3	**Preventive and protective measures** The risks assessed on JRK2 will reveal the preventative and protective measures necessary to comply with relevant regulations and will be recorded on JRK3.
Form JRK 4	**Safe work system** The risk assessment procedure will reveal work activities which require the design and implementation of a safe work system which will be recorded on JRK4.

Health surveillance

Regulation 5

Alternative 1: Workplace

The following surveillance requirement is revealed by the risk assessed process.

Zone	
Activity	
Persons affected	
Surveillance	From: To:
Person carrying out surveillance	
Disease or condition	

Form JRK 5	**Health surveillance record** Form JRK 5 records the individual circumstances where Health Surveillance has been implemented

Alternative 2: Work activities

Work activity
Location
Persons affected
Surveillance
Person carrying out surveillance
Disease of condition

Procedures for situations of serious and imminent danger

Regulation 7

Risk assessment has revealed the following situations of serious or imminent danger which might occur:

Event	Affected area	Action to be taken
Fire	All areas	Evacuate
Gas escape	Plant room	Evacuate

Names of competent person	Duty zone	Appointment From To

Excavation procedure

- The fire alarm system signals the implementation of the evacuation procedure.
- An intermittent signal in your zone is an alert. Be ready to evacuate but do not do so until instructed by the duty zone competent person.
- A continuous signal in your zone indicates an emergency in that zone and you must evacuate immediately to the designated assembly point.
- Regular exercises will be held.

Employers sharing a workplace: co-operation and co-ordination arrangements

Regulation 9

Host employer	Address
Employers permanently sharing workplace	**Areas of common occupation**

Employers temporarily sharing workplace		**Areas of common occupation**
Employer		
Occupation		
Workplace tasks		
Competent person		
Time on site		

Health and safety co-ordinator	**Common risks assessed**
Name	1.
Organization	2.
	3.
Phone	4.
	5.

Employee capability and training

Regulation 11

- Employees work within their skills abilities and training.
- Employers must identify task-specific training needs and provide appropriate training.

Employee Name	Job Description	Induction Training	Training Provided
1.			
2.			
3.			
4.			

Employee Name	Job Description	Specific Training	Training Provided
1.			
2.			
3.			
4.			

Employee Name	Job Description	Refresher Training	Training Provided
1.			
2.			
3.			
4.			

Employee duties:

- Use equipment as trained.
- Do not use equipment for any other purpose.
- Inform employer of dangerous situations.
- Notify employer of health and safety shortcomings.

Standard forms for use in conjunction with policy document

JRK 1: Hazard identification form

Hazard: Potential to cause harm

Identified hazard	Zone	Persons affected
1.		
2.		
3.		

JRK 2: Risk assessment form

Risk: Likelihood potential will be realized

Hazard	Applies to		Risk arising from hazard	Severity		
	Safety	Health		High	Med.	Low

JRK 3: Preventative and protective measures

Risk	Preventive measure	Protective measure

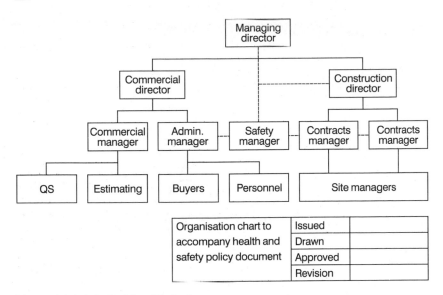

Figure A1.1 Safe Builders Limited: management structure

Appendix 2:
Worked example of a construction company health and safety policy

Subject: Safety procedures		Ref: 100
		Issue: 1
Distribution		Date 1/9/99
Prepared by	Approved	Page 1

Preface

This is the 1999 edition of the Safe Builders Limited Health and Safety Policy Manual. It is duly approved by the managing director and supersedes all previous editions.

Distribution is controlled. However, it is always available for review as required at the company's head office which is 18 Millers Yard, Mill Lane, Cambridge. Additional copies are available from the safety manager who will ensure that the circulation list is amended.

On no account must the manual or its contents be copied or disclosed to non-company personnel or organizations without prior written authority from the issuing office.

Aim, objectives and scope of this manual
The aim of this policy manual is to positively communicate our health, safety and environmental commitments to our employees and personnel associated with this company.

Our objectives are to eliminate harm by:

- continually improving the quality of safety and;
- creating a behaviour based safety culture, thus reducing financial and physical loss.

The criteria will contribute to the vision, mission and business performance of the company.

The scope defines:

- the company's health and safety documentation, organization, systems, procedures and responsibilities;
- how health and safety performance will be measured, monitored and reviewed.

Date of next review 1st September 2000

Subject: Safety procedures index		Ref: 101 Issue: 1
Distribution: Safety procedures manual holders		Date 1/9/99
Prepared by	Approved	Page 2

The following are authorized manual holders:

Manual number	Holder
01	Managing director
02	Commercial director
03	Construction director
04	Safety manager
05	Contracts manager
06	Contracts manager
07	Site manager
08	Site manager
09	Site manager
10	Site manager
11	Site manager
12	Administration manager
13	Personnel manager

Please note that the master manual is held by the safety manager whose name is . . . . and he can be contacted through the company's head Office which is at 18 Millers Yard, Mill Lane, Cambridge (Tel 01223 461633).

Please note that all amendments to the policy document will be carried out solely by the safety manager, whose responsibility it is to keep the manual up to date at all times.

Subject: Company policy for health and safety general statement		Ref: 102 Issue: 1
Distribution		Date 1/9/99
Prepared by	Approved	Page 3

1. General policy statement

It is the policy of this company to:

- Ensure the health, safety and welfare at work of all its employees or any other person authorized to be on company business.
- Maintain the company's plant and systems of work in a safe condition and without risk to health as far as is reasonably practicable.
- Provide and maintain means of entrance and exit from the company's workplaces that are safe and without risks

2. Responsibility

The managing director shall have overall and final responsibility for health, safety and welfare on all of the company's sites and in its head office and shall ensure that there is an effective policy which is implemented and integrated into the business strategy.

The safety manager is responsible to the managing director for ensuring the company is advised of changes in legislation or corporate policies and is administered through the administration manager.

3. Implementation

Specific policies, allocation of responsibilities and particular arrangements for the implementation of this general policy statement are set out in the safety procedures Manual.

4. Review

This policy will be under constant review for updating with change in legislation or company policies and formally reviewed annually to ensure all aspects of the documents are correct and current.

Managing director   Dated:

Subject: Organizational responsibilities – directors		Ref: 103 Issue: 1
Distribution: Safety procedures manual holders		Date 1/9/99
Prepared by	Approved	Page 4

1. Policy

The following designated company officials are responsible for ensuring implementation of the company's safety policy under specific headings.

2. Construction director

The construction director is responsible to the managing director of Safe Builders Limited for the implementation of the company's safety policy.

3. Board of directors

Directors are responsible to the construction director for ensuing that the company's safety policy and procedures are effectively imple-mented by ensuring that every manager within their function is familiar with and carries out their responsibilities as laid out in the Safety Procedures Manual.

4. Review

Whilst procedures should be reviewed constantly, a complete and formal review will take place each September on the instructions of the construction director. The construction director will delegate this task to the safety manager.

5. Document retention

Registers and records maintained in pursuance of the company's statutory duties under health and safety law are preserved for a minimum period of two years from the last date of entry. Accident books are preserved for a minimum of three years from the last date of entry. These registers and records are stored in the safety manager's office

Subject: Organizational responsibilities – safety manager		Ref: 104 Issue: 1
Distribution: Safety procedures manual holders		Date 1/9/99
Prepared by	Approved	Page 5

1. Policy

The safety manager reports to the managing director through the construction director and has responsibility for ensuring that the company complies with all necessary statutory safety regulations and company corporate policies and procedures.

2. Audits

2.1 Ensure compliance with company safety procedures by means of safety Audits.

2.2 Safety procedures – These must be amended and updated as necessary.

2.3 Training – Ensure all necessary safety training is carried out in conjunction with the administration manager and personnel manager.

2.4 Reports and statistics – Provide reports and statistics as required by the company, the Health and Safety Executive or other government departments.

2.5 Security – Recommend actions to be taken for the improvement of site and security and ensure compliance with existing policies.

2.6 Liaison – Liaise with external bodies, governmental, industrial or private as may be necessary.

Subject: Organizational responsibilities – contracts managers		Ref: 105 Issue: 1
Distribution: Safety procedures manual holders		Date 1/9/99
Prepared by	Approved	Page 6

Policy

Contracts managers are responsible to the construction director for the implementation of the company's safety policy and ensuring that all site managers within their region are familiar with and carry out their responsibilities as laid down in the Safety Procedures Manual. In addition they will undertake the following specific tasks:

1. Day-to-day co-ordination on health and safety arrangements.

2. Carry out appropriate liaison with clients and other contractors.

3. Undertake any accident investigation necessary.

4. Ensure that all prescribed notices and safety signs are erected and maintained on all of the company's sites.

5. Reporting accidents and notifiable incidents to the Health and Safety Executive.

6. Oversee and monitor the necessary statutory inspections.

7. In conjunction with site managers, carry out risk assessments and compile method statements for all work activities which require them.

8. Be aware of any environmental pollution or noise/nuisance requirements and identify and implement appropriate measures to address the necessary regulations.

Subject: Organizational responsibilities – site managers		Ref: 106 Issue: 1
Distribution: Safety procedures manual holders		Date 1/9/99
Prepared by	Approved	Page 7

Policy

Site managers are responsible to their contracts manager for the implementation of the company's safety policy and ensuring that all operatives and employees under their control are familiar with and carry out their responsibilities as laid down in the Safety Procedures Manual. Specifically, site managers will undertake the following:

1. Machinery safety

Ensure that all machinery, plant and equipment, in particular safety devices, in use on their sites are fully maintained and operational.

2. Risk assessment

In conjunction with the contracts managers carry out all necessary risk assessments leading to the compilation of method statements on all work which requires it.

3. Safe work systems

Ensure that safe work systems are identified for all elements of work which require them and ensure that the identified systems are fully implemented.

4. Training

It is the responsibility of all site managers to ensure that all those under their control including other supervisors and all operatives and employees both of the company and of their subcontractors are fully trained in safe operating procedures of all plant and equipment and machinery and safe work systems.

5. Personal protective equipment (PPE)

Site managers must ensure the appropriate PPE is readily available to all personnel and activity encourage its use where not mandatory.

6. New equipment

Site managers must ensure that no new plant, equipment or machinery is used without the approval of the safety officer and without ensuring that all those who may use it have received appropriate training.

Subject: Organizational responsibilities – site managers		Ref: 106 Cont Issue: 1
Distribution: Safety procedures manual holders		Date 1/9/99
Prepared by	Approved	Page 8

7. Housekeeping

It is incumbent on all site managers to uphold the high standards of housekeeping throughout the company.

8. Accidents

All accidents and dangerous occurrences are to be reported in the first instance to the contracts manager and are then to be investigated in accordance with company policy

9. Liaison with management on health and safety matters

Site managers are to ensure that the views of operatives and employees including subcontractors are sought on a regular basis and passed on to the company's management.

10. Co-ordination

Day-to-day co-ordination with clients and contractors is the responsibility of the site managers and the co-ordination arrangements are to be communicated to the contracts manager.

11. Visitors and contractors

Site managers must ensure that all visitors and contractors and subcontractors abide by company regulations as laid down in the safety policy. In particular, this extends to admitting only authorized persons to any of the company's projects and building sites.

12. Authorization procedures for visitors and operatives

The site manager should be responsible for ensure that all visitors book in and out of the Visitors' Book which is to be kept on site at all times and that they abide by standard procedures for visitors.

The site manager is also responsible for ensuring that all subcontractors employees have received adequate induction training and are given a copy of the health and safety plan for the project in so far as it applies to their work. The site manager will ensure that they sign a written acknowledgement confirming receipt of the health and safety plan.

If subcontractors' operatives fail to work in accordance with the health and safety plan, the site manager will immediately inform his contracts manager and the operative's employer, who will be notified in writing and requested to take immediate steps to compel the operative to comply or to replace him with another operative.

Subject: Organizational responsibilities – subcontractors	Ref: 107
	Issue: 1
Distribution: Safety procedures manual holders	Date 1/9/99

Prepared by	Approved	Page 9

1. Policy
All subcontractors will be expected to comply with the company policy on health and safety and must ensure their own company policy is made available on site whilst work is carried out.

2. Relevant laws
All work must be carried out in accordance with the relevant statutory provisions and taking into account the safety of others on the site and the general public.

3. Scaffolding and cradles used
Scaffolding and cradles used by subcontractors' employees must be inspected by their employer or a competent person appointed by their employer to ensure that it is erected and maintained in accordance with the regulations and approved codes of practice.

4. Subcontractors
Subcontractors' employees are not permitted to alter anything provided for their use or interfere with any plant or equipment on the site unless authorized by the company's site manager.

Subject: Organizational responsibilities – operatives		Ref: 108 Issue: 1
Distribution: Safety procedures manual holders		Date 1/9/99
Prepared by	Approved	Page 10

The responsibilities and duties of operatives are:

1. Read and understand the company policy on health and safety and work in accordance with its requirements.

2. Use the correct tools and equipment for the job.

3. Wear safety footwear at all times and use all protective clothing and safety equipment provided.

4. Keep tools and equipment in good order.

5. Report immediately to the site manager any defects in plant or equipment.

6. Work in a safe manner at all times. Do not endanger yourself or others by taking unnecessary risks. If possible remove site hazards yourself.

7. Do not use unsuitable plant and equipment for work that it was not intended for.

8. Do not use plant or equipment for which you have received no training or have no experience in the use of.

9. If you know of any particular hazards, warn other employees about them.

10. Do not engage in horseplay or practical jokes on site.

11. Report any person seen abusing any facilities, particularly the welfare facilities, to the site manager.

12. Report any injury to yourself which results from an accident at work even if the injury does not stop you from working.

13. Report any damage to plant or equipment.

14. Suggest safer methods of working.

15. If you are taking any medication which could affect your ability to carry out a work task inform your site manager.

Appendix 3:
References and sources

HSE publications

All available from: HSE Books
                    PO Box 1999
                    Sudbury
                    Suffolk CO10 6FS

HSE public enquiry point: 0114–289–2345 (phone)
                          0114–289–2333 (facsimile)

Health and Safety at Work etc. Act 1994. Management of Health and Safety at Work Regulations 1992 and Approved Code of Practice.

Managing Construction for Health and Safety Construction (Design and Management) Regulations 1994 and Approved Code of Practice.

The Construction (Health, Safety and Welfare) Regulations 1996 Statutory Instrument SI 1996 No. 1592.

Reporting of Industrial Diseases and Dangerous Occurrence Regulations 1996.

Workplace (Health, Safety and Welfare) Regulations 1992 and Approved Code of Practice.

Provision and Use of Work Equipment Regulation 1992 and Guidance on Regulations.

Control of Substances Hazardous to Health Regulations 1988 and 1994, and Approved Codes of Practice.
Manual Handling Operations Regulations 1992.
Personal Protective Equipment of Work Regulations 1992.
Health and Safety (Display Screen Equipment) Regulations 1992.
Noise at Work Regulations 1989.
Control of Asbestos at Work Regulations 1987.
Control of Lead at Work Regulations 1980.
Successful Health and Safety Management HS(G) 65.
Five Steps to Risk Assessment.
A Guide to Risk Assessment Requirements IND(G) 218(L).
A Guide to the Health and Safety at Work etc. Act 1974: Guidance on the Act.
The Health and Safety System in Great Britain.
An Introduction to Health and Safety, INDG 259.
The cost of Accidents at Work HS(G) 96.
Everyone's Guide to RIDDOR 95 HSE 31.
Fire Safety in Construction Work HSG 168.
Various HSE information sheets in the 'Construction Sheet' Series, 1–100, and in particular Sheets 39–44 dealing with CDM.
Designing for Health and Safety in Construction (1995). A Guide for Designers to the CDM Regulations 1994.
Review of Health and Safety Regulations. Main Report May 1994.
Managing Contractors: A Guide for Employers.

Other laws or regulations

The Building Act 1984.
The Building Regulations 1991.
The Building (Approved Inspectors etc.) Regulations 1985.
The Building (Prescribed Fees) Regulations 1994.

General references and sources

Association of Consulting Engineers' Terms and Conditions of Appointment. Thomas Telford.
Audit Commission (1997) *Rome was not Built in a Day: A Management Handbook on Getting Value for Money from Capital Programmes and Construction Projects*. Audit Commission Publications Book Point Limited, 39 Milton Park, Abingdon, Oxfordshire OX14 4TD. ISBN 1 86240 0164

Banwell Report (Ministry of Public Buildings and Works) (1964) The Placing and Managing of Contracts for Building and Civil Engineering. Chairman Sir H. Banwell. HMSO, London.

Bennett, J. and Jayes, S. (1995) *Trusting the Team: The Best Practice Guide to Partnering in Construction*. Reading Construction Forum.

Chapman, (1994) Risk Analysis in Project Management, ed. J. Raftery. E. & F.N Spoon.

CIRIA Report 117: Value by Competition.

CIRIA Report 172 (1998) Practical Guidance for Clients and Clients Agents under the CDM Regulations.

CIRIA Report 173 (1998) Practical Guidance for Planning Supervisors.

Constructing the Team. Sir Michael Latham (Latham Report, 1994). HMSO, ISBN 011 752994X.

Construction Procurement by Government: An Efficiency Unit Scrutiny (1995) (Levene Report). HMSO, ISBN 011 4301417

CIRIA Special Publication 132 (1996) Quality Management in Construction: Survey of Experiences with BS 5750. Report of Key Findings.

Code of Estimating Practice (1997) 6th edn, Chartered Institute of Building, Longman, ISBN 0562 30279X.

Code of Practice for Project Management for Construction and Development (1996) Chartered Institute of Building, Longman, ISBN 0 582 27680 2.

Construction Industry Board (1997) Work Group Reports 1–14 inclusive. Thomas Telford

Construction Site Safety Notes (1994) GE 700. Construction Publication; Construction Industry Training Board.

Cox, A. and Thompson, I. (1997) *Contracting for Business Success*, Thomas Telford, ISBN 07277 26005.

DETR (1997) Construction Research Business Plan. Department of the Environment, Transport and the Regions.

Dewis, M. (1995) *Tolley's Health and Safety at Work Handbook*. Tolley Publishing Co. Ltd. ISBN 0 85459 945 2.

Dickens, C. *Hard Times* Penguin. ISBN, 014 043398 8.

Economic Development Committee for Building (1967) Action of Banwell Report.

HM Treasury CUP Guides (1998):
- No. 1: Essential Requirements for Construction Procurement
- No. 2: Value for Money in Construction Procurement
- No. 3: Appointment of Consultants and Contractors.

Institution of Chemical Engineers' Model Forms of Contract for Process Plant. Redwood Books (Red, Blue and Green books) Guide, ISBN 0 85295 371 2.

ISO 9000 (1994) British Standards Institution.

Joint Contracts Tribunal publications by RIBA:
- Standard Form of Building Contract 1980 edition
- Practice No. 27
- Amendment 14
- Amendment 18
- Practice Note CD/2.

Law of Contract (1986) Cheshire Fifoot and Furmston, Butterworth, ISBN 406 56536 8.

Masterman, J.W.E. (1992) *An Introduction to Building Procurement Systems*, EFN Spon, ISBN 0 419 17720 5.

Turner, A. (1990) *Building Procurement*, MacMillan Press, ISBN 0 333 52286 9.

Oakland, J.S. (1989) *Total Quality Management*. Butterworth–Heinemann. ISBN 0 7506 0993 1.

O'Reilly, J.J.N. (1987) *Better Briefing Means Better Buildings*, BRE, BR 95, ISBN 0 85125 213 3.

Partnering in the Public Sector (1997). European Construction Institute. ISBN 1 873844 34 4.

Powell-Smith, V. and Billington, M.J. (1995) *The Building Regulations Explained and Illustrated*. Blackwell Science ISBN 0 632 039337

Rethinking Construction: The Report of the Construction Task Force. (Egan Report, 1998) HMSO, ISBN 1 85112 0947.

RIBA Publications (1996) Architect's Guide to Job Administration under the CDM Regulations 1994, ISBN 1 85946 018 6.

Simon Report (Central Council for Building and Works) (1944) The Placing and Managing of Building Contracts. Chairman Sir E. Simon. HMSO, London.

Standard Form of Agreement for the Appointment of an Architect (SFA 1992). RIBA Publications.

Scheme for Construction Contracts (England and Wales) Regulations 1998.

Setting New Standards. A Strategy for Government Procurement (1995) CM 2840. HMSO, ISBN 0 10 128402 0.

Shingo, S. *Poka-Yoke. Improving Product Quality by Preventing Defects* Productivity Press. ISBN 0 915299 31 3.

The Engineering and Construction Contract, Guidance Document to the Suite of Forms (1995) Thomas Telford, ISBN 0 7277 1941 6.

The Lifting Operation and Lifting Equipment Regulations 1998.

Towill (1997) Construction Business Review, 6

Wood Report (Building Economic Development Committee (1975) The Public Client and the Construction Industries. Chairman Sir H. Wood. NEDO, London.

Index